# Organic and Inorganic Chemistry

# Organic and Inorganic Chemistry

Dakota Rooney

Larsen & Keller
www.larsen-keller.com

Organic and Inorganic Chemistry
Dakota Rooney
ISBN: 978-1-64172-632-0 (Hardback)



Larsen & Keller
Published by Larsen and Keller Education,
5 Penn Plaza,
19th Floor,
New York, NY 10001, USA

**Cataloging-in-Publication Data**

Organic and inorganic chemistry / Dakota Rooney.
p. cm.
Includes bibliographical references and index.
ISBN 978-1-64172-632-0
1. Chemistry, Organic. 2. Chemistry, Inorganic. 3. Chemistry. I. Rooney, Dakota.
QD251.3 .C44 2022
547--dc23



For more information regarding Larsen and Keller Education and its products, please visit the publisher's website www.larsen-keller.com

# Table of Contents

# Preface

This book has been written, keeping in view that students want more practical information. Thus, my aim has been to make it as comprehensive as possible for the readers. I would like to extend my thanks to my family and co-workers for their knowledge, support and encouragement all along.

Organic and inorganic chemistry are sub-disciplines of chemistry that study organic and inorganic compounds respectively. Organic chemistry studies the structure, properties and reactions of organic compounds. Such compounds contain carbon in covalent bonding. It is important to study their structure to determine their chemical composition and formula. This branch of chemistry studies the physical and chemical properties of organic compounds and evaluates their chemical reactivity to understand their behavior. Inorganic chemistry focuses on the synthesis and behavior of inorganic and organometallic compounds. Inorganic compounds are derived from nature as minerals. This book is a valuable compilation of topics, ranging from the basic to the most complex theories and principles in the field of organic and inorganic chemistry. Some of the diverse topics covered in this book address the varied branches that fall under this category. It will provide comprehensive knowledge to the readers.

A brief description of the chapters is provided below for further understanding:

Chapter - Introduction to Organic and Inorganic Chemistry

The branch of science which studies the composition of matter along with their properties and reactions is known as chemistry. It is broadly divided into two categories, namely, inorganic chemistry and organic chemistry. This chapter discusses in detail the theories and methodologies related to these two branches of chemistry as well as their importance.

Chapter - Organic Compounds

Organic compounds generally refer to the compounds that contain carbon atoms. Some of the common organic compounds are hydrocarbons, acetylacetone and formaldehyde. Hydrocarbons are divided into two broad classes, namely aliphatic and aromatic compounds. This chapter has been carefully written to provide an easy understanding of the varied facets of organic compounds.

Chapter - Organic Reactions

The chemical reactions which involve organic compounds are known as organic reactions. Some of the major types of organic reactions are substitution reactions, addition reactions, elimination reactions, rearrangement reactions and oxidation-reduction reactions. The chapter closely examines these reactions to provide an extensive understanding of the subject.

Chapter - Inorganic Compounds

A compound which lacks a carbon-hydrogen bond is generally known as an inorganic compound. Some of the common inorganic compounds are ammonium bicarbonate, sodium fluoride, beryllium oxide, calcium sulfate and sodium amide. The topics elaborated in this chapter will help in gaining a better perspective about these different inorganic compounds.

Chapter - Inorganic Reactions

The reactions which involve inorganic compounds are known as inorganic reactions. Some of the major types of inorganic reactions are combination reactions, decomposition reactions, single-replacement reactions and double-replacement reactions. This chapter has been carefully written to provide an easy understanding of these types of reactions.

Chapter - Organometallic Chemistry

The branch of chemistry which studies the chemical compounds which contain bonds between carbon and metal atoms is known as organometallic chemistry. Most organometallic compounds are used as catalysts in a number of industries. The diverse applications of organometallic compounds have been thoroughly discussed in this chapter.

**Dakota Rooney**

# 1

# Introduction to Organic and Inorganic Chemistry

The branch of science which studies the composition of matter along with their properties and reactions is known as chemistry. It is broadly divided into two categories, namely, inorganic chemistry and organic chemistry. This chapter discusses in detail the theories and methodologies related to these two branches of chemistry as well as their importance.

## Organic Chemistry

Organic chemistry is the area of chemistry that involves the study of carbon and its compounds. Carbon is now known to form a seemingly unlimited number of compounds. The uses of organic compounds impact our lives daily in medicine, agriculture, and general life.

In theory organic chemistry may have its beginnings with the big bang when the components of ammonia, nitrogen, carbon dioxide and methane combined to form amino acids, an experiment that has been verified in the laboratory. Organic chemicals were used in ancient times by Romans and Egyptians as dyes, medicines and poisons from natural sources, but the chemical composition of the substances was unknown.

In the 16th century organic compounds were isolated from nature in the pure state and analytical methods were developed for determination of elemental composition.

Scientists believed that organic chemicals found in nature contained a special "vital force" that directed their natural synthesis, and therefore, it would be impossible to accomplish a laboratory synthesis of the chemicals. Fortunately, later in the century Frederich Wöhler discovered that urea, a natural component in urine, could be synthesized in the laboratory by heating ammonium cyanate. His discovery meant that the natural "vital force" was not required to synthesis organic compounds, and paved the way for many chemists to synthesize organic compounds.

By the middle of the nineteenth century many advances had been made into the discovery, analysis and synthesis of many new organic compounds. Understanding about the structures of organic chemistry began with a theory of bonding called valence theory.

Organic chemistry developed into a productive and exciting science in the nineteenth century. Many new synthetic methods, reaction mechanisms, analytical techniques and structural theories have been developed. Toward the end of the century much of the knowledge of organic chemistry has been expanded to the study of biological systems such as proteins and DNA. Volumes of information are published monthly in journals, books and electronic media about organic and biological chemistry.

## The Chemical Bond

### Atomic Theory

The atomic theory of electrons began in the early 1900s and gained acceptance around 1926 after Heisenberg and Schroedinger found mathematical solutions to the electronic energy levels found in atoms, the field is now called quantum mechanics.

Electrons exist in energy levels that surround the nucleus of the atom. The energy of these levels increases as they get farther from the nucleus. The energy levels are called shells, and within these shells are other energy levels, called subshells or orbitals, that contain up to two electrons. The calculations from atomic theory give the following results for electron energy and orbitals. The results for the first two energy levels (shells 1 and 2) are the most important for bonding in organic chemistry.

Orbitals

| Shell | s | p | d | f | Total Electrons Possible |
|---|---|---|---|---|---|
| 1 | 1 | | | | 2 |
| 2 | 2 | 3 | | | 8 |
| 3 | 3 | 3 | 5 | | 18 |
| 4 | 1 | 3 | 5 | 7 | 32 |

- Energy level 1 contains up to two electrons in a spherical orbital called a 1s orbital.
- Energy level 2 contains up to eight electrons; two in an 2s-orbital and two in each of three orbitals designated as 2p-orbitals. The p-orbitals have a barbell type shape and are aligned along the x, y, and z axes. They are thus called the px, py, and pz orbitals.

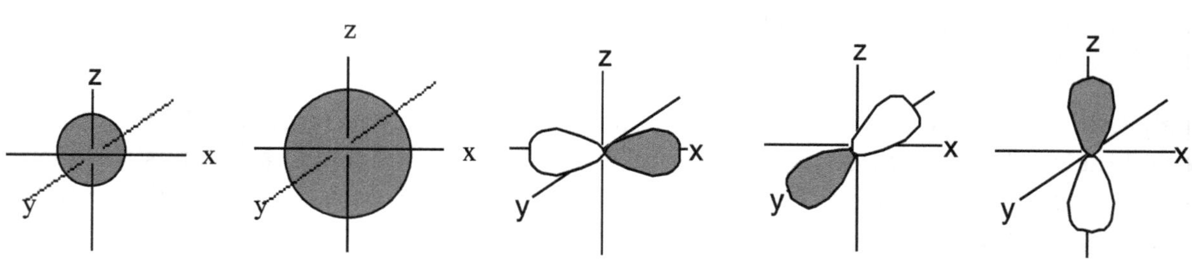


1s orbital 2s orbital 2px orbital 2py orbital 2pz orbital

- Energy level 3 contains up to eighteen electrons, two electrons in a 3s orbital, six electrons in the three 3p orbitals, and ten electrons in the five 3d orbitals.
- Energy level 4 contains up to thirty-two electrons, two electrons in a 4s-orbital, six electrons in the three 4p-orbitals, ten electrons in the five 4d-orbitals, and fourteen electrons in the seven 4f-orbitals.

Electrons fill the lower energy levels first until all of the electrons are used (Aufbau Principle). An element contains the number of electrons equal to its atomic number. For the first and second row elements the electron configurations are relatively simple.

| Element (atomic number) | Electron Configuration |
|---|---|
| H (1) | $1s^1$ (1st shell, s orbital, one electron) |
| He (2) | $1s^2$ |
| Li (3) | $1s^2, 2s^1$ |
| Be (4) | $1s^2, 2s^2$ |
| B (5) | $1s^2, 2s^2, 2p^1$ |
| C (6) | $1s^2, 2s^2, 2p^2$ |
| N (7) | $1s^2, 2s^2, 2p^3$ |
| O (8) | $1s^2, 2s^2, 2p^4$ |
| F (9) | $1s^2, 2s^2, 2p^5$ |
| Ne (10) | $1s^2, 2s^2, 2p^6$ (inert, completely filled) |

## Electronegativity

Electronegativity is the ability of an atom to attract electrons to itself, and generally increases as one moves from left to the right across the periodic table.

least Electronegative    Li < Be < B < C < N < O < F    most electronegative

Electronegativity also increases as we go from the bottom to the top of a column in the periodic table.

least Electronegative    I < Br < Cl < F    most electronegative

Elements that easily lose electrons and attain a positive charge are called electropositive elements. Alkali metals are electropositive elements.

## Bonding

Atoms can become bonded with each other, and their electronic structure governs the type of bond formed. The main two types of bonds that are formed are called ionic and covalent.

## Ionic Bond

Ionic bonding is important between atoms of vastly different electronegativity. The bond results from one atom giving up an electron while another atom accepts the electron. Both atoms attain a stable nobel gas configuration.

In the compound lithium fluoride, the $2s^1$ electron of lithium is transferred to the $2p^5$ orbital of fluorine. The lithium atom gives up an electron to form the positively charged lithium cation with $1s^2$, $2s^0$ configuration, and the fluorine atom receives an electron to form a fluoride anion with $1s^2$, $2s^2$, $2p^6$ configuration. Thus the outer energy levels of both ions are completely filled. The ions are held together by the electrostatic attraction of the positive and negative ions.

| Li | + | F | $\longrightarrow$ | $Li^+$ | + | $F^-$ |
|---|---|---|---|---|---|---|
| $1s^2$ | | $1s^2$ | | $1s^2$ | | $1s^2$ |
| $2s^1$ | | $2s^2$ $2p^5$ | | $2s^0$ | | $2s^2$ $2p^6$ |

## Covalent Bond

A covalent bond is formed by a sharing of two electrons by two atoms. A hydrogen atom possessing the $1s^1$ electron joins with another hydrogen atom with its $1s^1$ configuration. The two atoms form a covalent bond with two electrons by sharing their electrons.

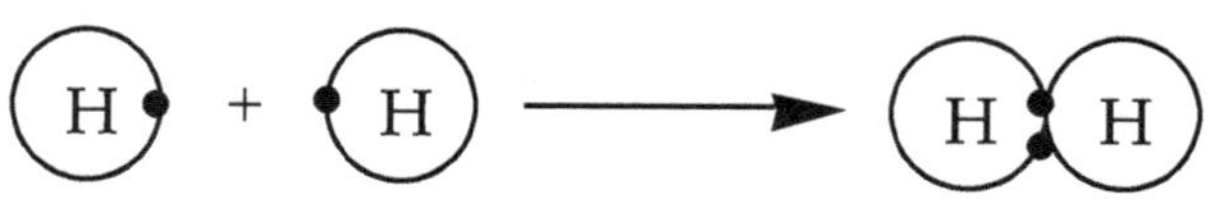


In hydrogen fluoride, HF, the hydrogen 1s electron is shared with a 2p5 electron in fluorine ($1s^2$, $2s^2$, $2p^5$), and the molecule is now held together by a covalent bond. In this case, the fluorine atom is much more electronegative than the hydrogen atom and the electrons in the bond tend to stay closer to the fluorine atom. This is called a polar covalent bond, and the atoms possess a small partial charge denoted by the Greek δ symbol.

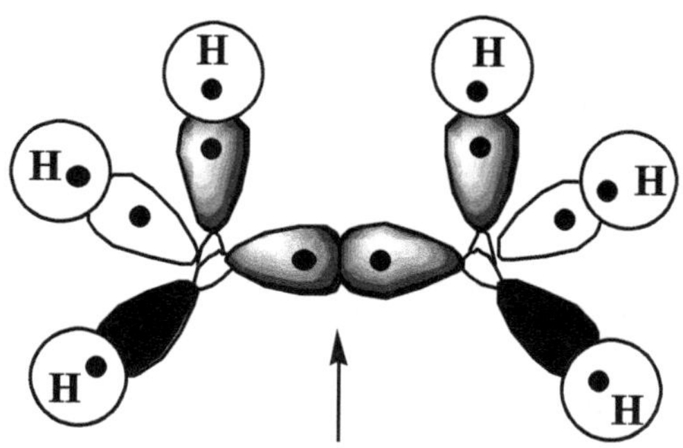


Carbon-carbon sigma bond

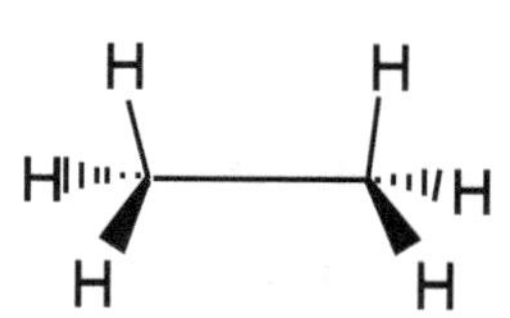


Ethane

## Bonding in Carbon Compounds

The property of carbon that makes it unique is its ability to form bonds with itself and therefore allows a large number of organic chemicals with many diverse properties. Carbon has the property of forming single, double and triple bonds with itself and with other atoms. This multiple bond ability allows carbon compounds to have a variety of shapes. In all carbon compounds, carbon forms four bonds. The types of bonds used by the carbon atom are known as sigma (σ) and pi (π) bonds. Different combinations of these bonds lead to carbon single bonds, double bonds and triple bonds.

## The Carbon-Hydrogen Single Bond-The Sigma (σ) Bond

σ By far most of the bonds in carbon compounds are covalent bonds found commonly in the carbon-hydrogen single bond. In carbon ($1s^2$, $2s^2$, $2p^2$) one of the electrons of the $2s^2$ orbital is promoted to the third $2p^o$ orbital. The s and three p orbitals hybridize to form four new orbitals of equal energy called $sp^3$ hybrid orbitals. The electrons in the four sp3 hybridized orbitals bond by overlap with the $1s^1$ hydrogen orbital. The single covalent bond is called a sigma (σ) bond. The sp3 bonds arrange themselves as far from each other as possible, the shape of a molecule of methane, $CH_4$, is tetrahedral with 109.5° bond angles.

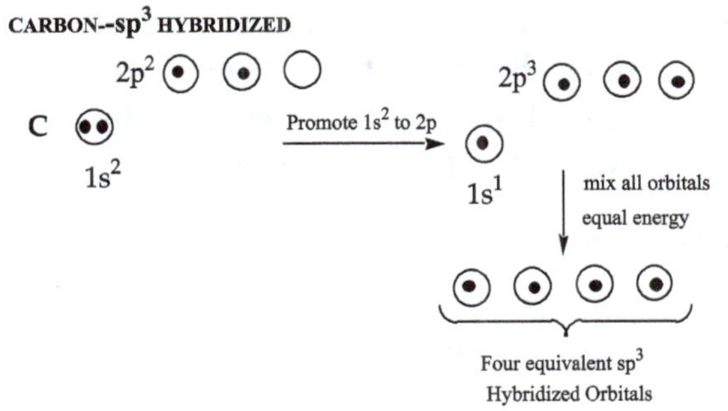


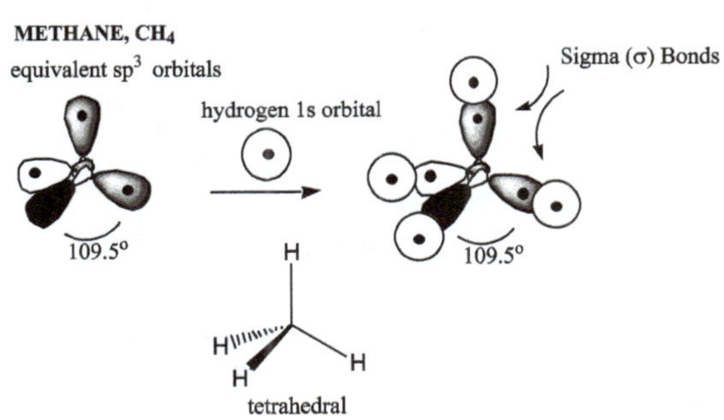


The unique property of carbon that differentiates it from the other elements and allows the formation of so many different organic compounds is the ability of carbon to bond with itself through covalent bonding. Thus, addition of another carbon atom to methane results in ethane which has covalent sigma bonds to the hydrogen atoms and a covalent sigma bond between the carbon atoms. Addition of more carbon atoms leads to many more compounds.

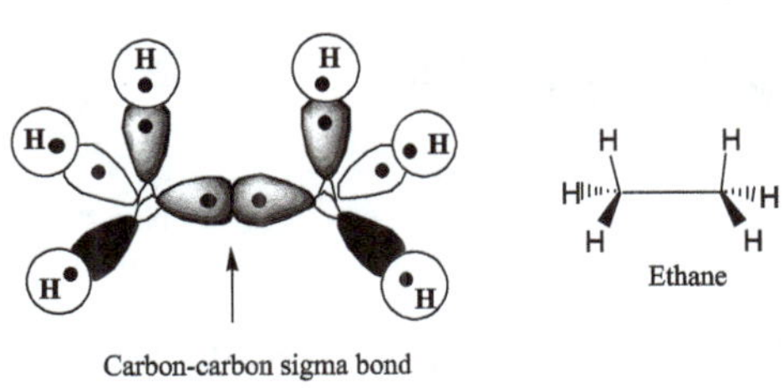


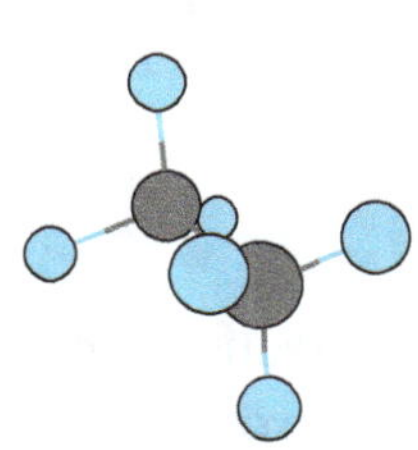

## The Carbon-Carbon Double Bond-The Pi ($\pi$) Bond

Carbon forms a wide variety of compounds that contain carbon bonded to another carbon with a double bond between the two atoms. These compounds are classified as alkenes (older naming calls them olefins). The orbital model below explains the carbon-carbon double bond. The carbon electron configuration shows one's electron being promoted to a p orbital. But now only three orbitals are mixed, a s orbital and two p orbital, that are called $sp^2$ hybrid orbitals and are used to form single bonds (sigma bonds). The p orbital contains one electron.

The combination of two of the $sp^2$ hybridized carbon atom leads to two carbon atoms being joined by overlap of $sp^2$ orbitals to form a C-C single bond, and the side-to-side overlap of the p orbitals to form another bond known as a pi ($\pi$) bond. In the molecule of ethene shown below. There are a total of 5 sigma bonds and one pi bond. As a result of the bonding in an ethene, the molecule is planar with m bond angles of 120° and a C=C bond length that is longer than the C-H bond length.

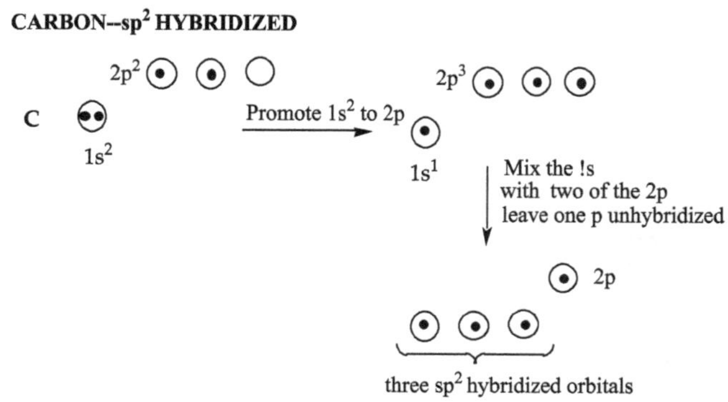


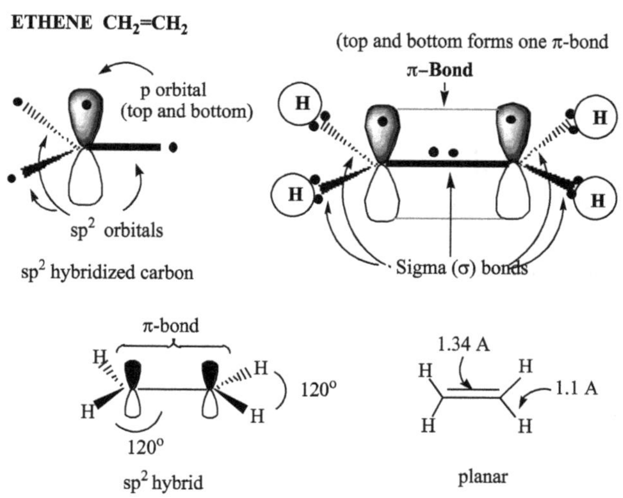


When carbon forms a bond with an electronegative atom such as oxygen, nitrogen, sulfur or a halogen, the bond is a polar covalent bond with the electrons of the bond residing closer to the electronegative atom.

## The Carbon-Carbon Triple Bond

Another type of bond that carbon forms with itself is the triple bond found in a class of compounds called alkynes. After promotion of the 2s electron to a 2p orbital, one's orbital mixes with one p orbital to give two hybrid sp orbitals. The two remaining p orbitals are used to make p bonds. Thus the carbon is bound by a sigma bond to hydrogen from one of the sp hybrid orbital, to the other carbon atom by a sigma bond from one of the sp hybrid orbitals, and the two carbon atoms are bound by two pi bonds from side-to-side overlap of the two p orbitals. The sp hybrid orbitals position themselves 180° apart and thus a molecule of ethyne is linear with the hydrogen atoms 180° apart.

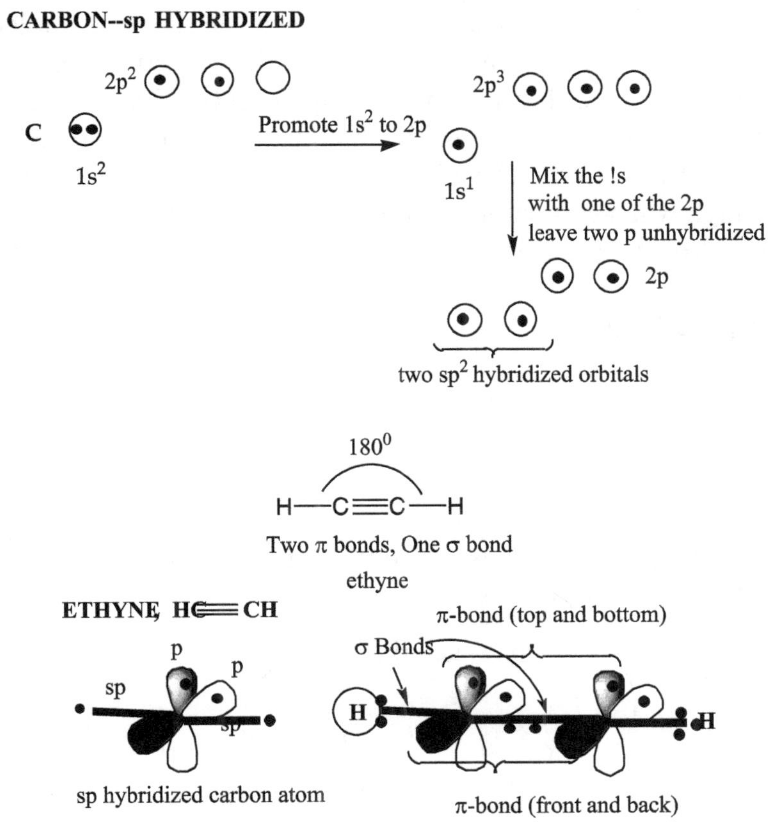


## Polar Covalent Bonds in Carbon

Carbon forms single, double and triple bonds with elements other than carbon. The atoms involved in the bonding are usually oxygen, nitrogen, sulfur and the halogens. These elements are more electronegative than carbon and thus attract the electrons to themselves. The bonds are therefore polar covalent bonds. Bonds that contain a separation of charge possess a dipole moment, a property that contributes to the overall polarity of the molecule.

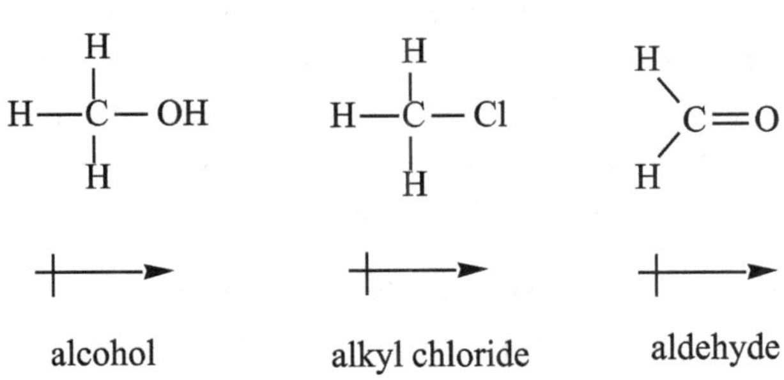

## Hydrogen Bonds and Bond Polarity

The bonds O-H, N-H and F-H are highly polar covalent bonds because the electronegative draws electrons away from the hydrogen atom. In every case the hydrogen atom has a partial positive charge.

$O^{\delta-}$-$H^{\delta+}$ $N^{\delta-}$-$H^{\delta+}$ $F^{\delta-}$-$H^{\delta+}$

A result of molecules having these highly polarized bonds with the hydrogen atom partly positive in nature, the hydrogen atom is attracted to the basic site in other molecules, such as the non-bonding electrons on oxygen and nitrogen (non-bonding electrons are electrons belong to an element that complete the octet but do not participate in bonding). This attraction is called hydrogen bonding and is useful for explaining high boiling points and high melting points of fairly low mass molecules. Thus hydrogen bonding in water explains why the compound with only three atoms boils relatively high when compared with other molecules of similar mass. Extra energy is required to break the hydrogen bonds during the boiling process.

## Hydrogen Bonding in Water

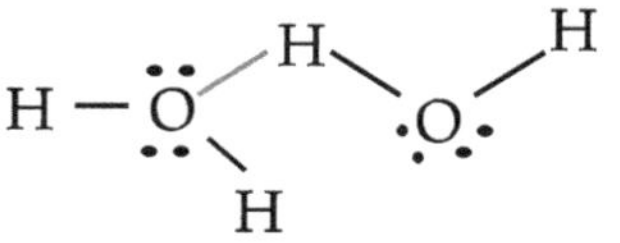

In organic compounds hydrogen bonding is very important for describing the boiling and solubility characteristics of alcohols and acids.

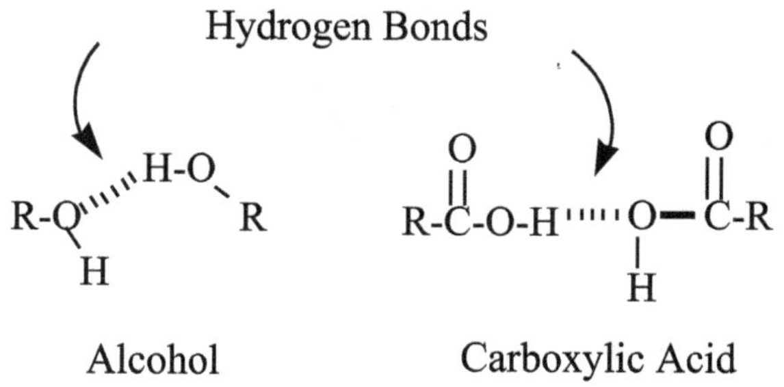


R stands for organic function.

## Organic Structures

At first the study of organic chemistry involves understanding the energy and shapes of simple organic structures. Then the study progresses to more complicated three-dimensional structures and finally to the investigation of reactions of organic molecules. We use the theories of bonding and the facts about acidic or basic properties of the molecules to understand the overall picture of what is happening. But the overall goal is to improve mankind through the development of new and better processes to be applied in medicine, agriculture and biomedical science that is to make life better. Thus we must be able to study difficult molecules too.

As you proceed through organic chemistry you will write many different structures, and the writing gets easier with practice. But in the beginning it can be confusing. Always remember that carbon has four bonds.

In writing an alkane such as pentane which has five carbon atoms. We could write it by just showing carbon atoms, but it does not show carbon with four bonds. More correctly it needs all of its hydrogen atoms, but writing all the hydrogen atoms for compounds becomes cumbersome. Thus we write condensed structures as shown below. With some experience, we can write pentane as the line structure with the ends and the points representing carbon atoms, and we know that the hydrogens are there without writing them.

C-C-C-C-C

Pentane with no H atoms

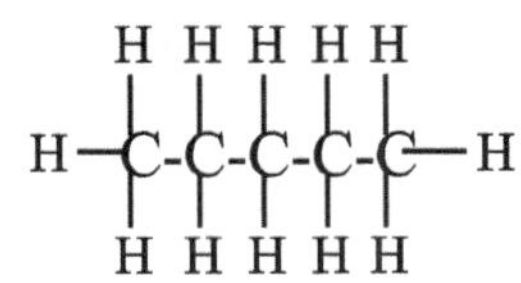

Pentane with all H atoms

$CH_3$-$CH_2$-$CH_2$-$CH_2$-$CH_3$

Pentane condensed structure

$CH_3CH_2CH_2CH_2CH_3$

Pentane completely condensed structure

Pentane

Pentane Advanced condensed Structure.

Carbon compounds also exist in rings, called cyclic compounds. Below is the compound called cyclohexane written several different ways. Even in the most condensed structures you must realize that all the carbon atoms and all the hydrogen atoms are present even though the C's and H's are not written. These shorthand methods of writing structures are very helpful in speeding up the writing of equations, and thus understanding them is necessary.

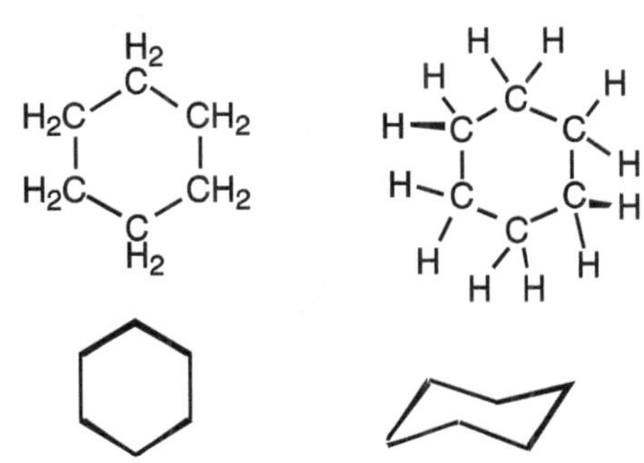

The structures of organic chemicals can be very complicated and very interesting. Sometime they are so complex that condensed structures are the only reasonable presentation. But as always, the carbon atom has four bonds. Several advanced organic structures are shown below:

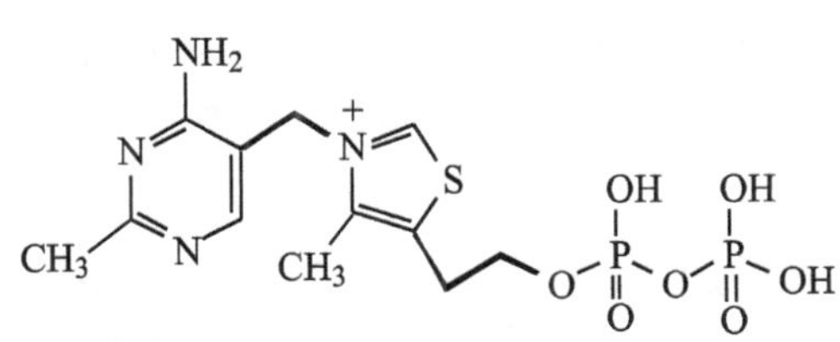

Thiamin diphosphate Vitamin B1 (grains, liver, pork).

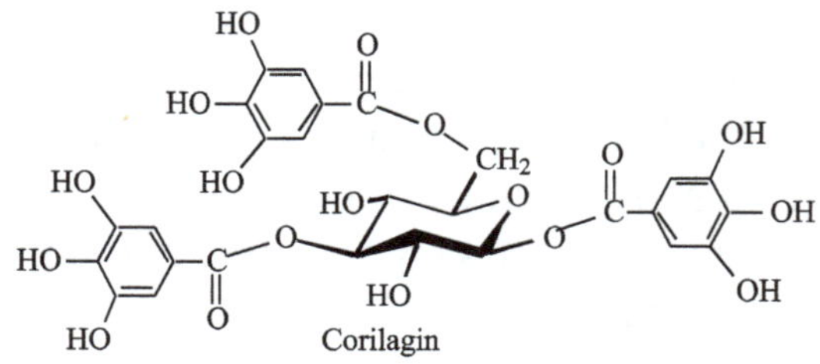

Corilagin

A molecule in the tannin family found in cranberries prevents bacteria from attaching to the kidneys.

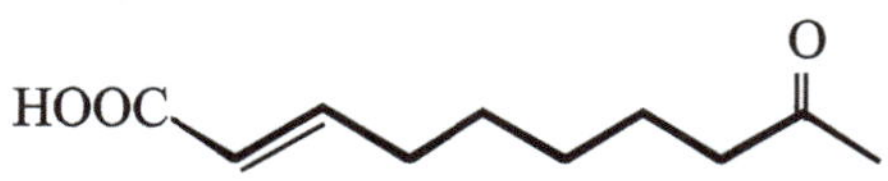

sex pheromone of bees

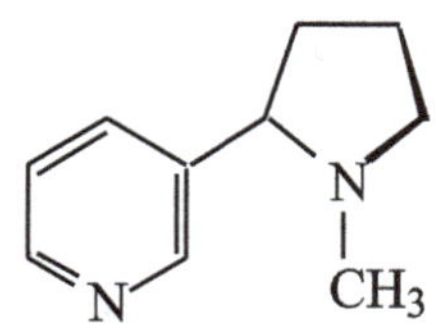

Nicotine
Addictive component
of tobacco.

## Importance of Organic Chemistry

1. Medicine: Medicine is the main applications of organic compounds. Though not all but many drugs are made of organic substances. Like antibiotics, anticancer drugs, painkillers, anti-depressant, anesthetics, cardiac drugs, etc.

In medicine, organic chemistry finds its use in three ways like the drugs, to study the disease and for diagnosis.

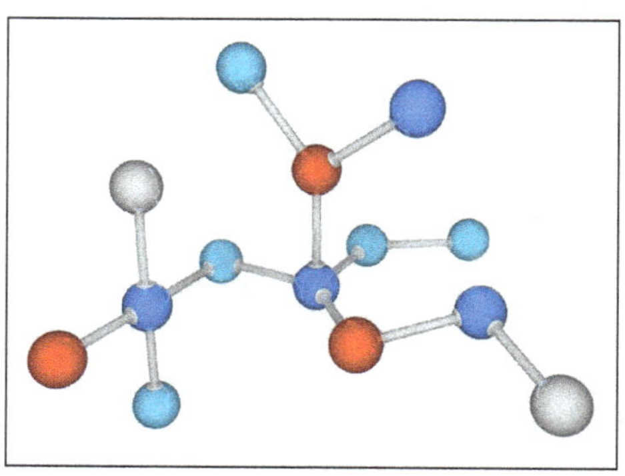

Cool design.

- Drugs to cure disease: As said before many drugs used for the treatment of diseases are made of organic compounds. Hence they are water-insoluble, bitter and also easily movable in the body tissues.

The drug delivery to the deeper-most parts of the body tissues needs enhancement of lipid solubility. This can be achieved by the change in the organic ration, i.e., the amount of carbon content will give the desired effect.

Hence the organic chemistry studies the chemistry of the drug and tries to enhance their efficiency, tissue reach and their safety. They are made safe by minimizing their toxicity through the process of metabolism.

Further, organic compounds having the same chemical structure have varying effect in the body due to stereoisomerism. The 'Cis' and 'Trans' isomers play a different role. If you notice anti-Parkinson medicine L-DOPA or an antibiotic Levofloxacin, you can find both drugs having L-configuration (-). They are Levo isomers of the same substance, but Levo form is more effective than the Dextro forms (+).

- Pathophysiology of the diseases: The study of disease is possible by the knowledge of biochemistry and organic chemistry. Most diseases in humans have some course or pathway before death ensues.

Ex: In gout, there is a disturbance in the purine metabolism; Purines are important moieties of DNA and RNA molecules. The uric acid formed during purine metabolism does not further breakdown into urea and gets accumulated. The urea crystals get accumulated in the smaller joints causing gout. This can be studied by checking the blood levels of uric acid with that of a normal person.

Similarly, in infections, there is damage to some biochemical components in the body, for example in malaria. The parasites damage the hemoglobin content of the blood. In this case, the hemoglobin levels go down. This can be recognized by the change in organic functional groups in between the normal and diseased patients.

This change in the organic components helps us to study the course and severity of the disease.

- To diagnose the disease: Here the organic chemistry uses some diagnosing aids to detect the organic part of the disturbed substance.

In diabetics, there is increased sugar levels and in severe cases even the ketone levels. Sugars have aldehyde groups (CHO) and ketones (C=O) groups. These groups are the targets in the analysis. Higher concentration of these groups during estimation, the more is the sugar levels and vice-versa. So organic chemistry helps in disease diagnosis by checking for the organic functional group of the disturbed substance in the body.

In heart patients, the cholesterol levels from blood are estimated.

2. Food: Food materials are entirely composed of carbon compounds viz. carbohydrates (CHO), proteins (NH2-CH-COOH), and fats (CH-COO-CH). Even vitamins are organic in nature. Based on the study of the body requirements during pregnancy, disease condition, body fitness, experts advise the use of proteins, vitamins and other substituents. For example folic acid is recommended during pregnancy to minimize the chances of anemia in mother. For the heart patient, doctors advice for minimal consumption of fat diet. And for those interested in bodybuilding they advise for high consumption of protein-rich diets.

Even beverages made of ethyl alcohol like the beer, wines, vodka have organic matter in them. The knowledge of organic chemistry helps regulates the flavor, quality and storage of these beverages.

3. Textiles and clothing: Cloth is made of cotton, wool, silk, polyester, etc. All these materials are organic. This chemistry helps in the study of textiles material, their improvement for durability, color and cleaning process.

4. Cleansing agents: In industries and labs, organic solvents are widely used for cleaning. For example in drug extraction from plants, the fatty matter from the pulp is removed using petroleum ether.

At home and other places of living, we use phenols and other agents to clean the floor and walls. These sanitizing agents work on the principle of organic chemistry to remove the dirt and also kill the microbes.

Thus organic chemistry through its knowledge of polarity, solubility, partition factors uses solvents for better use.

5. Sterilizing agents: Most of the sterilizing agents and disinfectants like phenol, formaldehyde, etc. are carbon compounds. Due to their properties like solubility, pH they can kill microbes and even human body cells.

They kill the bacteria and other microbes by either dissolving the microbe cell wall or damaging the protein layers. Their efficiency is enhanced by making small tweaks in the compound chemistry.

Besides these solvents, there are gases like ethylene oxide which are used for sterilization of drugs and manufactured substances.

6. Analytic materials: Chemical substances like drugs, pesticides, cosmetics, are tested as a part of quality control. This testing is done by using different types of titrations, chromatography techniques, and spectrophotometry.  Here the reagents like acids, or bases, reducing and oxidizing agents used are organic. Further, the endpoint indicators in the titration are also organic compounds.

7. Valuables: Diamonds, graphite, petroleum. Interestingly the carbon compounds are found to be highly valuable, durable and hardest in the world.

Diamond and graphite are both pure carbon alone compound without any other elements inside. They are both highly used and expensive. Their properties are studies in organic chemistry. Petroleum is the other most valued resources on the earth for fuels needs in the world. These petroleum products are further diversified for various uses. And petroleum is one of the factors which influence the world economy.

## Other Applications and uses

Organic chemistry finds its application to a large extent in the field of medicine, pesticides, petroleum sector, textile, etc.

1. For analysis: Not all organic substances are soluble in water. So they can be analyzed by non-aqueous titration. For this, they use organic solvents like pyridine, methanol, acetone, etc.

Further other methods like chromatography, spectroscopy (infrared spectroscopy) also employ organic solvents for analysis. This analysis helps to test the given compound for its purity, quantity, etc.

2. For synthesis: Organic chemistry helps us synthesis many compounds which are needed on a large scale. For example, we find a drug molecule in nature by chance or a bi-product of some reaction. Then that drug molecule can be synthesized by knowledge of organic chemistry for large-scale use.

So we see many drug manufacturing companies filing for patents to manufacture the same drug in their methods. The same molecule can be made by using different steps of the reaction. But pharma companies try to make it by an organic reaction which is inexpensive for better profits.

3. Identification: This another area where the importance of organic chemistry plays a major role. Most substances either extracted from plants, animals are identified by the use of specific tests. These tests employ organic compounds and related techniques.

4. For better molecules: If a molecule is already used for a long time. It can be replaced by similar molecules with a slight change in its chemistry. This change is aimed at more effective performance. This is possible by substituting some organic functional groups.

Thus the importance of organic chemistry is immense in human life.

# Inorganic Chemistry

Inorganic chemistry is defined as the study of the chemistry of materials from non-biological origins. Typically, this refers to materials not containing carbon-hydrogen bonds, including metals, salts, and minerals. Inorganic chemistry is used to study and develop catalysts, coatings, fuels, surfactants, materials, superconductors, and drugs. Important chemical reactions in inorganic chemistry include double displacement reactions, acid-base reactions, and redox reactions.

In contrast, chemistry of compounds that contain C-H bonds is called organic chemistry. The organometallic compounds overlap both organic and inorganic chemistry. Organometallic compounds typically include a metal directly bonded to a carbon atom.

The first man-made inorganic compound of commercial significance to be synthesized was ammonium nitrate. Ammonium nitrate was made using the Haber process, for use as a soil fertilizer.

## Properties of Inorganic Compounds

Because the class of inorganic compounds is vast, it's difficult to generalize their properties. However, many inorganics are ionic compounds, containing cations and anions joined by ionic bonds. Classes of these salts include oxide, halides, sulfates, and carbonates. Another way to classify inorganic compounds is as main group compounds, coordination compounds, transition metal compounds, cluster compounds, organometallic compounds, solid state compounds, and bioinorganic compounds.

Many inorganic compounds are poor electrical and thermal conductors as solids, have high melting points, and readily assume crystalline structures. Some are soluble in water, while others are not. Usually the positive and negative electrical charges balance out to form neutral compounds. Inorganic chemicals are common in nature as minerals and electrolytes.

## Importance of Inorganic Chemistry

Inorganic Chemistry is not an isolated branch of chemistry. This core science is fully integrated with other areas of chemistry such as organic, physical and analytical chemistry. It deals with the chemistry of all non-organic compounds, and mainly involves the chemistry of metals and especially transition metals. These elements play a crucial role in industrial catalytic processes that are required to produce substances and new materials at a rate far exceeding that of natural chemical reactions. Such catalytic processes can take place in solution (homogeneous catalysis) or on the surface of solid materials (heterogeneous catalysis) and usually involve transition metal elements.

However, such elements also play a crucial role in biological processes (so-called bioinorganic chemistry) where metallo-enzymes can activate small molecules like $O_2$, $H_2O_2$, NO, $H_2$, CO and $CO_2$, which then act as oxygen transfer reagents, participate in hydration processes, function as messenger molecules or form essential components of redox biology. Furthermore, on the basis of such catalytic reactivity, many metallo-drugs have been developed for the treatment of cancer, arthritis, multiple sclerosis and other autoimmune diseases.

For example, the anti-tumor drug cisplatin, *cis*-$[Pt(NH_3)_2(Cl)_2]^{2+}$, was already developed in the sixties. In the meantime, it is known that it slowly aquates to *cis*-$[Pt(NH_3)_2(H_2O)Cl]^+$, which can bind to DNA bases in the cell and cause apoptosis of cancer cells. Also in the area of atmospheric chemistry, transition metal catalysts are successfully used for the treatment of gaseous effluents from coal fired power plants, for selective catalyzed reduction (SCR technology) of $NO_x$ in exhaust gases from cars and trucks, and pollution caused by heating with low quality coal during winter times.

A huge challenge to inorganic chemists is the drive to optimize existing and develop new technology that will improve the performance of catalysts to save energy and aim for sustainable developments. In this respect, the clarification of the underlying reaction mechanisms in order to understand the underlying chemical processes, whether of industrial, environmental or biological significance, is of utmost importance to the whole world in order to tackle threatening climate changes and severe pollution in densely populated cities. Here inorganic chemistry can indeed have an impact on the quality of life and the wellbeing of the increasing population of the world.

# 2
# Organic Compounds

Organic compounds generally refer to the compounds that contain carbon atoms. Some of the common organic compounds are hydrocarbons, acetylacetone and formaldehyde. Hydrocarbons are divided into two broad classes, namely aliphatic and aromatic compounds. This chapter has been carefully written to provide an easy understanding of the varied facets of organic compounds.

Organic compounds always contain carbon along with other elements that are needed for living organisms to function. Carbon is the key element because it has four electrons in an outer electron shell that can hold eight electrons. As a result, it can form many types of bonds with other carbon atoms and elements such as hydrogen, oxygen and nitrogen. Hydrocarbons and proteins are good examples of organic molecules that can form long chains and complex structures. The organic compounds made up of these molecules are the basis for chemical reactions in the cells of plants and animals – reactions that provide the energy for finding food, for reproduction and for all the other processes necessary for life.

## Characteristics of Organic Compounds

The four types of organic compounds are hydrocarbons, lipids, proteins and nucleic acids, and they perform different functions in a living cell. While many organic compounds are not polar molecules and therefore don't dissolve well in the water of a cell, they often dissolve in other organic compounds. For example, while carbohydrates such as sugar are slightly polar and dissolve in water, fats do not. But fats dissolve in other organic solvents such as ethers. When in solution, the four types of organic molecules interact and form new compounds as they come into contact within living tissue.

Organic compounds range from simple substances with molecules made up of a few atoms of only two elements to long, complex polymers whose molecules include many elements. Hydrocarbons for example, are made up of only carbon and hydrogen. They can form simple molecules or long chains of atoms and are used for cell structure and as basic building blocks for more complex molecules.

## Carbohydrates

Carbohydrates are organic compounds that contain only carbon, hydrogen, and oxygen. They are the most common of the four major types of organic compounds. There are thousands of different carbohydrates, but they all consist of one or smaller units called monosaccharides.

## Monosaccharides and Disaccharides

The general formula for a monosaccharide is:

$$(CH_2O)_n$$

Where $n$ can be any number greater than two. For example, if $n$ is 6, then the formula can be written:

$$C_6H_{12}O_6$$

This is the formula for the monosaccharide glucose. Another monosaccharide, fructose, has the same chemical formula as glucose, but the atoms are arranged differently. Molecules with the same chemical formula but with atoms in a different arrangement are called isomers. Compare the glucose and fructose molecules in Figure below. Can you identify their differences? The only differences are the positions of some of the atoms. These differences affect the properties of the two monosaccharides.

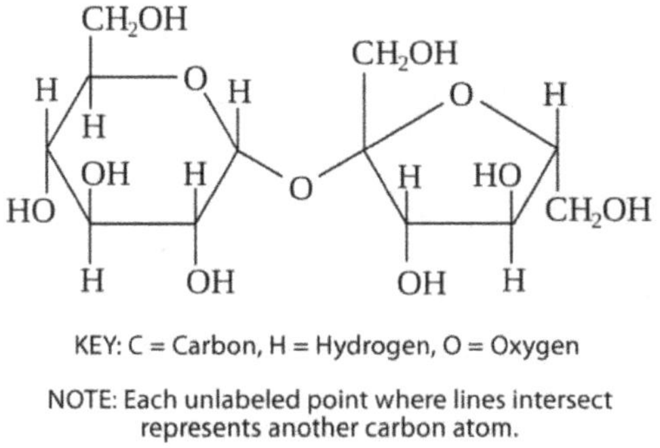


Figure: Sucrose Molecule. This sucrose molecule is a disaccharide. It is made up of two monosaccharides: Glucose on the left and fructose on the right.

If two monosaccharides bond together, they form a carbohydrate called a disaccharide. An example of a disaccharide is sucrose (table sugar), which consists of the monosaccharides glucose and fructose. Monosaccharides and disaccharides are also called simple sugars. They provide the major source of energy to living cells.

## Polysaccharides

If more than two monosaccharides bond together, they form a carbohydrate called a polysaccharide. A polysaccharide may contain anywhere from a few monosaccharides to several thousand monosaccharides. Polysaccharides are also called complex carbohydrates. Their main functions are to store energy and form structural tissues. Examples of several polysaccharides and their roles are listed in table.

## Complex Carbohydrates

| Complex Carbohydrate | Function | Organism |
|---|---|---|
| Amylose | Stores energy | Plants |
| Glycogen | Stores energy | Animals |

| Cellulose | Forms cell walls | Plants |
|---|---|---|
| Chitin | Forms external skeleton | Some animals |

These complex carbohydrates play important roles in living organisms.

## Lipids

Lipids are organic compounds that contain mainly carbon, hydrogen, and oxygen. They include substances such as fats and oils. Lipid molecules consist of fatty acids, with or without additional molecules. Fatty acids are organic compounds that have the general formula $CH_3(CH_2)_nCOOH$, where $n$ usually ranges from 2 to 28 and is always an even number.

## Saturated and Unsaturated Fatty Acids

Fatty acids can be saturated or unsaturated. The term saturated refers to the placement of hydrogen atoms around the carbon atoms. In a saturated fatty acid, all the carbon atoms (other than the carbon in the -COOH group) are bonded to as many hydrogen atoms as possible (usually two hydrogens). Saturated fatty acids do not contain any other groups except -COOH. This is why they form straight chains, as shown in figure. Because of this structure, saturated fatty acids can be packed together very tightly. This allows organisms to store chemical energy very densely. The fatty tissues of animals contain mainly saturated fatty acids.

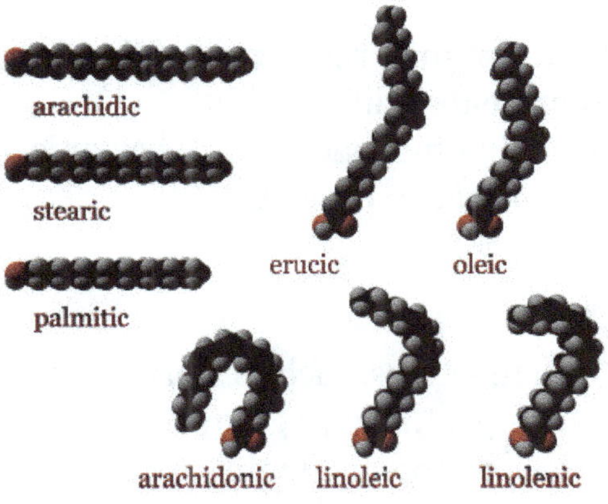


Figure: Saturated and Unsaturated Fatty Acids. Saturated fatty acids include arachidic, stearic, and palmitic fatty acids, shown on the left in this figure. Unsaturated fatty acids include all the other fatty acids in the figure. Notice how all the unsaturated fatty acids have bent chains, whereas the saturated fatty acids have straight chains.

In an unsaturated fatty acid, some carbon atoms are not bonded to as many hydrogen atoms as possible. This is because they are bonded to one or more additional groups, including double and triple bonds between carbons. Wherever these other groups bind with carbon, they cause the chain to bend - they do not form straight chains. This gives unsaturated fatty acids different properties than saturated fatty acids. For example, unsaturated fatty acids are liquids at room temperature whereas saturated fatty acids are solids. Unsaturated fatty acids are found mainly in plants, especially in fatty tissues such as nuts and seeds.

Unsaturated fatty acids occur naturally in the bent shapes shown in figure. However, unsaturated fatty acids can be artificially manufactured to have straight chains like saturated fatty acids.

Called trans fatty acids, these synthetic lipids were commonly added to foods, until it was found that they increased the risk for certain health problems. Many food manufacturers no longer use trans fatty acids for this reason.

## Types of Lipids

Lipids may consist of fatty acids alone or in combination with other compounds. Several types of lipids consist of fatty acids combined with a molecule of alcohol:

- Triglycerides are the main form of stored energy in animals. This type of lipid is commonly called fat. A triglyceride is shown in figure below.
- Phospholipids are a major component of the membranes surrounding the cells of all organisms.
- Steroids (or sterols) have several functions. The sterol cholesterol is an important part of cell membranes and plays other vital roles in the body. Other steroids are male and female sex hormones.

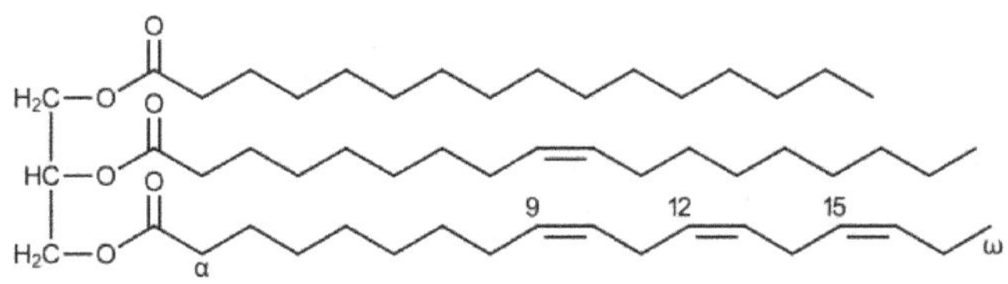

Triglyceride Molecule: The left part of this triglyceride molecule represents glycerol. Each of the three long chains on the right represents a different fatty acid. From top to bottom, the fatty acids are palmitic acid, oleic acid, and alpha-linolenic acid. The chemical formula for this triglyceride is $C_{55}H_{98}O_6$. Key: H=hydrogen, C=carbon, O=oxygen.

## Lipids and Diet

Humans need lipids for many vital functions, such as storing energy and forming cell membranes. Lipids can also supply cells with energy. In fact, a gram of lipids supplies more than twice as much energy as a gram of carbohydrates or proteins. Lipids are necessary in the diet for most of these functions. Although the human body can manufacture most of the lipids it needs, there are others, called essential fatty acids that must be consumed in food. Essential fatty acids include omega-3 and omega-6 fatty acids. Both of these fatty acids are needed for important biological processes, not just for energy. Although some lipids in the diet are essential, excess dietary lipids can be harmful. Because lipids are very high in energy, eating too many may lead to unhealthy weight gain. A high fat diet may also increase lipid levels in the blood. This, in turn, can increase the risk for health problems such as cardiovascular disease. The dietary lipids of most concern are saturated fatty acids, trans fats, and cholesterol. For example, cholesterol is the lipid mainly responsible for narrowing arteries and causing the disease atherosclerosis.

## Proteins

Proteins are organic compounds that contain carbon, hydrogen, oxygen, nitrogen, and, in some cases, sulfur. Proteins are made of smaller units called amino acids. There are 20 different common amino acids needed to make proteins. All amino acids have the same basic structure, which is shown in figure. Only the side chain (labelled R in the figure) differs from one amino acid to another. The variable side chain gives each amino acid unique properties. Proteins can differ from one another in the number and sequence (order) of amino acids. It is because of the side chains of the amino acids that proteins with different amino acid sequences have different shapes and different chemical properties.

Small proteins can contain just a few hundred amino acids. Yeast proteins average 466 amino acids. The largest known proteins are the titins, found in muscle, which are composed from almost 27,000 amino acids.

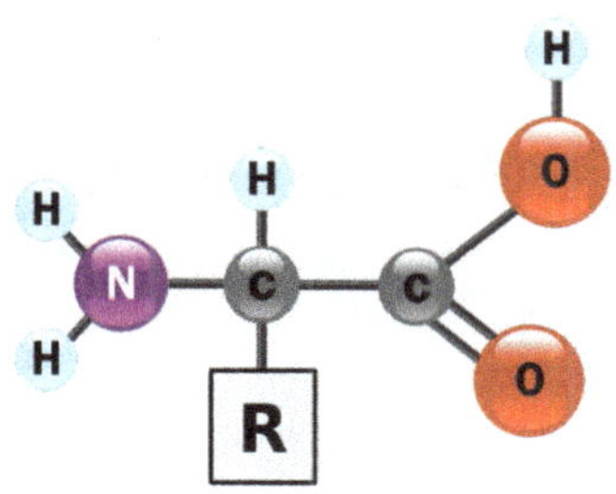


General Structure of Amino Acids. This model shows the general structure of all amino acids. Only the side chain, R, varies from one amino acid to another. For example, in the amino acid glycine, the side chain is simply hydrogen (H). In glutamic acid, in contrast, the side chain is $CH_2CH_2COOH$ . Variable side chains give amino acids acids different chemical properties. The order of amino acids, together with the properties of the amino acids, determines the shape of the protein, and the shape of the protein determines the function of the protein. Key: H = hydrogen, N = nitrogen, carbon, oxygen, R = variable side chain C=carbon, oxygen, R = variable side chain.

## Protein Structure

Amino acids can bond together to form short chains called peptides or longer chains called polypeptides. Polypeptides may have as few as 40 amino acids or as many as several thousand. A protein consists of one or more polypeptide chains. The sequence of amino acids in a protein's polypeptide chain(s) determines the overall structure and chemical properties of the protein. Primary protein structure is sequence of a chain of amino acids.

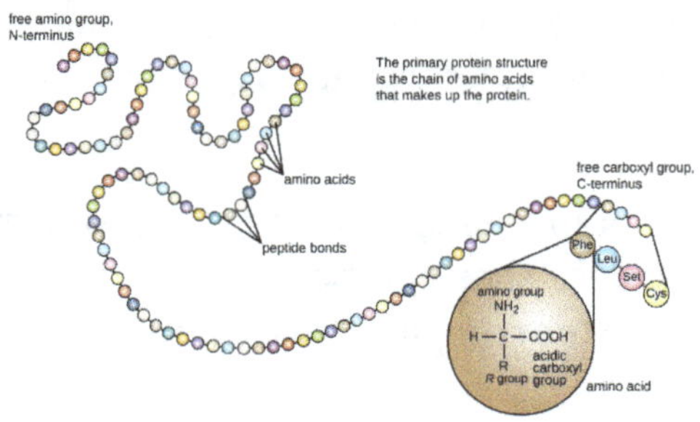


Figure: Polypeptide. This polypeptide is a chain made up of many linked amino acids.

The amino acid sequence is the primary structure of a protein. As explained in figure, a protein may have up to four levels of structure, from primary to quaternary. The complex structure of a protein allows it to carry out its biological functions.

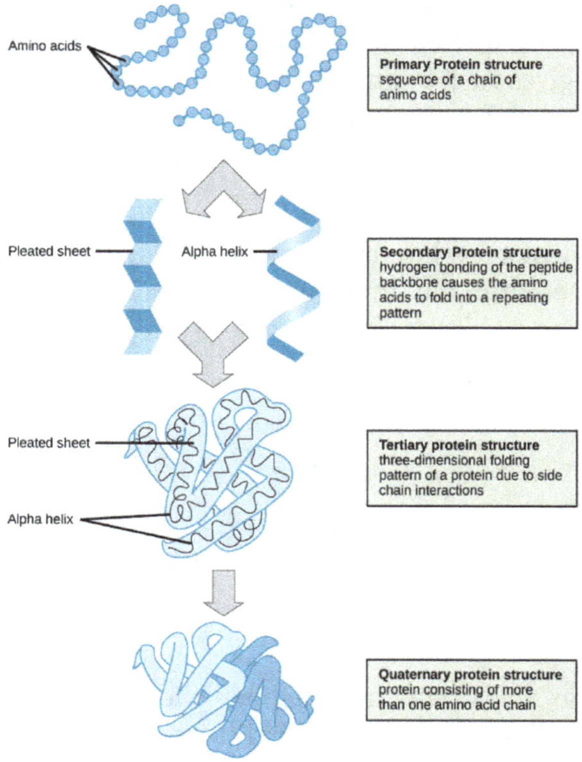


Primary protein structure is the sequence of amino acids in a single polypeptide. Secondary protein structure refers to internal shapes, such as alpha helices and beta sheets that a single polypeptide takes on due to bonds between atoms in different parts of the polypeptide. Tertiary protein structure is the overall three-dimensional shape of a protein consisting of one polypeptide. Quaternary protein structure is the shape of a protein consisting of two or more polypeptides.

## Functions of Proteins

Proteins are an essential part of all organisms. They play many roles in living things. Certain proteins provide scaffolding that maintains the shape of cells. Proteins also make up the majority of muscle tissues. Many proteins are enzymes that speed up chemical reactions in cells. Other proteins are antibodies. They bond to foreign substances in the body and target them for destruction. Still other proteins help carry messages or materials in and out of cells or around the body. For example, the blood protein hemoglobin bonds with oxygen and carries it from the lungs to cells throughout the body.

One of the most important traits of proteins, allowing them to carry out these functions, is their ability to bond with other molecules. They can bond with other molecules very specifically and tightly. This ability, in turn, is due to the complex and highly specific structure of protein molecules.

## Proteins and Diet

Proteins in the diet are necessary for life. Dietary proteins are broken down into their component amino acids when food is digested. Cells can then use the components to build new proteins. Humans are able to synthesize all but eight of the twenty common amino acids. These eight amino acids, called essential amino acids, must be consumed in foods. Like dietary carbohydrates and lipids, dietary proteins can also be broken down to provide cells with energy.

## Nucleic Acids

Nucleic acids are organic compounds that contain carbon, hydrogen, oxygen, nitrogen, and phosphorus. They are made of smaller units called nucleotides. Nucleic acids are named for the nucleus of the cell, where some of them are found. Nucleic acids are found not only in all living cells but also in viruses. Types of nucleic acids include deoxyribonucleic acid (DNA) and ribonucleic acid (RNA).

## Structure of Nucleic Acids

A nucleic acid consists of one or two chains of nucleotides held together by chemical bonds. Each individual nucleotide unit consists of three parts:

- A base (containing nitrogen).
- A sugar (ribose in RNA, deoxyribose in DNA).
- A phosphate group (containing phosphorus).

The sugar of one nucleotide binds to the phosphate group of the next nucleotide. Alternating sugars and phosphate groups form the backbone of a nucleotide chain, as shown in figure below. The bases, which are bound to the sugars, stick out at right angles from the backbone of the chain.

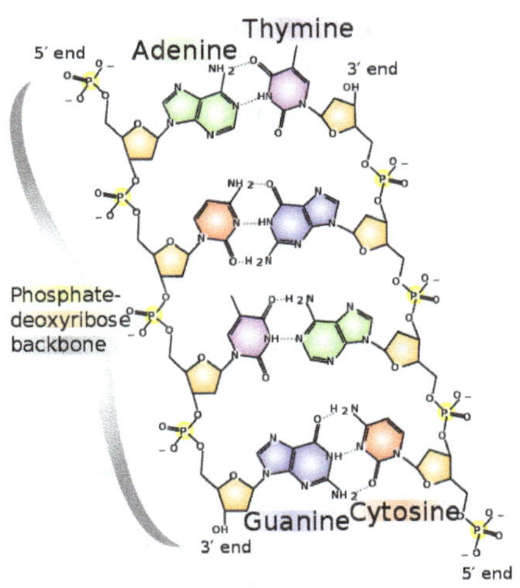


Figure: Part of a Nucleic Acid.

This small section of a nucleic acid shows how phosphate groups (yellow) and sugars (orange) alternate to form the backbone of a nucleotide chain. The bases that jut out to the side from the backbone are adenine (green), thymine (purple), cytosine (pink), and guanine (blue). Bonds between complementary bases, such as between adenine and thymine, hold the two chains of nucleotides together. These bonds, called hydrogen bonds.

RNA consists of a single chain of nucleotides, and DNA consists of two chains of nucleotides. Bonds form between the bases on the two chains of DNA and hold the chains together. There are four different types of bases in a nucleic acid molecule: cytosine, adenine, guanine, and either thymine (in DNA) or uracil (in RNA). Each type of base bonds with just one other type of base. Cytosine and guanine always bond together, and adenine and thymine (or uracil) always bond with one another. The pairs of bases that bond together are called complementary bases.

The binding of complementary bases allows DNA molecules to take their well-known shape, called a double helix. Figure below shows how two chains of nucleotides form a DNA double helix. A simplified double helix is illustrated in Figure below. It shows more clearly how the two chains are intertwined. The double helix shape forms naturally and is very strong. Being intertwined, the two chains are difficult to break apart. This is important given the fundamental role of DNA in all living organisms.

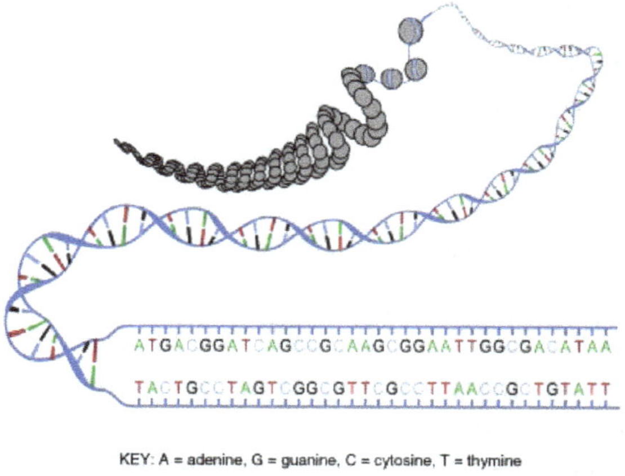


Double-Stranded Nucleic Acid. In this double-stranded nucleic acid, complementary bases (A and T, C and G) form bonds that hold the two nucleotide chains together in the shape of a double helix. Notice that A always bonds with T and C always bonds with G. These bonds help maintain the double helix shape of the molecule.

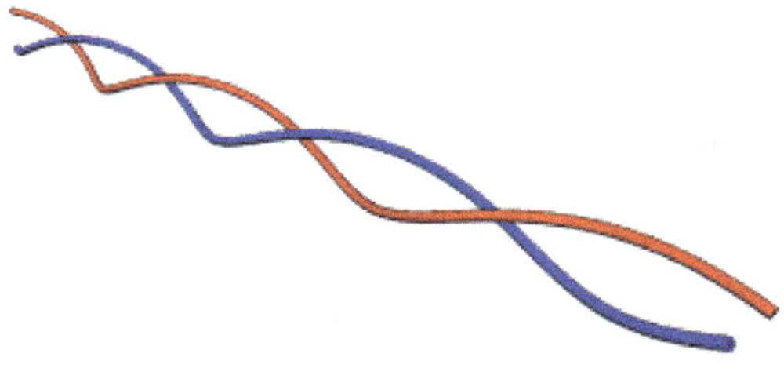

Figure: Simple Model of DNA. In this simple model of DNA, each line represents a nucleotide chain. The double helix shape forms when the two chains wrap around the same axis.

## Role of Nucleic Acids

The order of bases in nucleic acids is highly significant. The bases are like the letters of a four-letter alphabet. These "letters" can be combined to form "words." Groups of three bases form words of the genetic code. Each code word stands for a different amino acid. A series of many code words spells out the sequence of amino acids in a protein. In short, nucleic acids contain the information needed for cells to make proteins. This information is passed from a body cell to its daughter cells when the cell divides. It is also passed from parents to their offspring when organisms reproduce.

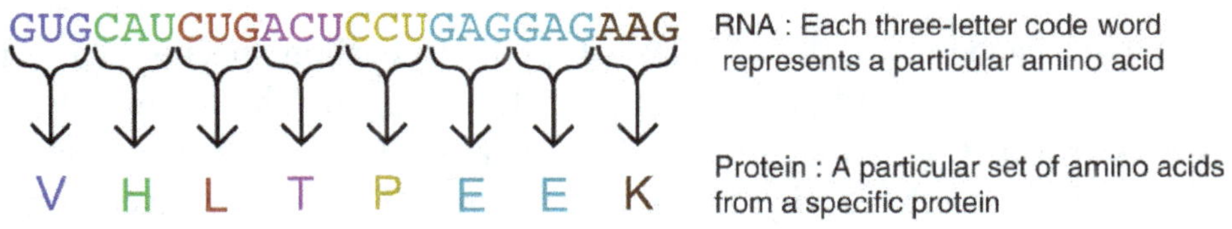


Figure: The letters G, U, C, and A stand for the bases in RNA. Each group of three bases makes up a code word, and each code word represents one amino acid (represented here by a single letter, such as V, H, or L). A string of code words specifies the sequence of amino acids in a protein.

### How RNA Codes for Proteins

DNA and RNA have different functions relating to the genetic code and proteins. Like a set of blueprints, DNA contains the genetic instructions for the correct sequence of amino acids in proteins. RNA uses the information in DNA to assemble the amino acids and make the proteins.

## Hydrocarbons

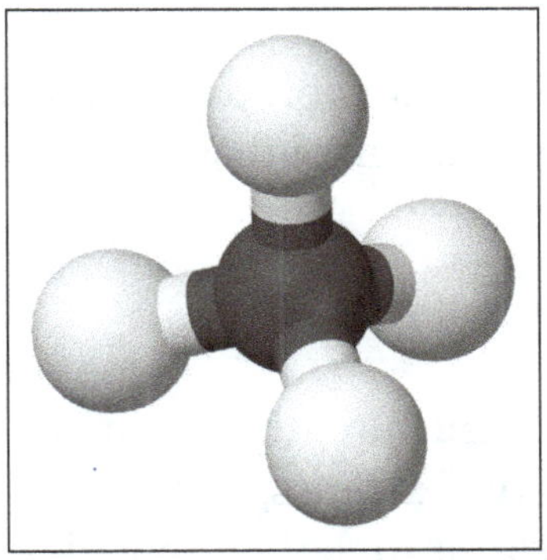

Ball-and-stick model of the methane molecule, $CH_4$. Methane is part of a homologous series known as the alkanes, which contain single bonds only.

In organic chemistry, a hydrocarbon is an organic compound consisting entirely of hydrogen and carbon. Hydrocarbons are examples of group 14 hydrides. Hydrocarbons from which one hydrogen atom has been removed are functional groups called hydrocarbyls. Because carbon has 4 electrons in its outermost shell (and because each covalent bond requires a donation of 1 electron, per atom, to the bond) carbon has exactly four bonds to make, and is only stable if all 4 of these bonds are used.

Aromatic hydrocarbons (arenes), alkanes, cycloalkanes and alkyne-based compounds are different types of hydrocarbons.

Most hydrocarbons found on Earth naturally occur in crude oil, where decomposed organic matter provides an abundance of carbon and hydrogen which, when bonded, can catenate to form seemingly limitless chains.

## Types of Hydrocarbons

As defined by IUPAC nomenclature of organic chemistry, the classifications for hydrocarbons are:

1. Saturated hydrocarbons are the simplest of the hydrocarbon species. They are composed entirely of single bonds and are saturated with hydrogen. The formula for acyclic saturated hydrocarbons (i.e., alkanes) is $C_nH_{2n+2}$. The most general form of saturated hydrocarbons is $C_nH_{2n+2(1-r)}$, where $r$ is the number of rings. Those with exactly one ring are the cycloalkanes. Saturated hydrocarbons are the basis of petroleum fuels and are found as either linear or branched species. Substitution reaction is their characteristics property (like chlorination reaction to form chloroform). Hydrocarbons with the same molecular formula but different structural formulae are called structural isomers. As given in the example of 3-methylhexane and its higher homologues, branched hydrocarbons can be chiral. Chiral saturated hydrocarbons constitute the side chains of biomolecules such as chlorophyll and tocopherol.

2. Unsaturated hydrocarbons have one or more double or triple bonds between carbon atoms. Those with double bond are called alkenes. Those with one double bond have the formula $C_nH_{2n}$ (assuming non-cyclic structures). Those containing triple bonds are called alkyne. Those with one triple bond have the formula $C_nH_{2n-2}$.

3. Aromatic hydrocarbons, also known as arenes, are hydrocarbons that have at least one aromatic ring.

Hydrocarbons can be gases (e.g. methane and propane), liquids (e.g. hexane and benzene), waxes or low melting solids (e.g. paraffin wax and naphthalene) or polymers (e.g. polyethylene, polypropylene and polystyrene).

## General Properties

Because of differences in molecular structure, the empirical formula remains different between hydrocarbons; in linear or "straight-run" alkanes, alkenes and alkynes, the amount of bonded hydrogen lessens in alkenes and alkynes due to the "self-bonding" or catenation of carbon preventing entire saturation of the hydrocarbon by the formation of double or triple bonds.

This inherent ability of hydrocarbons to bond to themselves is known as catenation, and allows hydrocarbons to form more complex molecules, such as cyclohexane, and in rarer cases, arenes such as benzene. This ability comes from the fact that the bond character between carbon atoms is entirely non-polar, in that the distribution of electrons between the two elements is somewhat even due to the same electronegativity values of the elements (~0.30), and does not result in the formation of an electrophile.

Generally, with catenation comes the loss of the total amount of bonded hydrocarbons and an increase in the amount of energy required for bond cleavage due to strain exerted upon the molecule; in molecules such as cyclohexane, this is referred to as ring strain, and occurs due to the "destabilized" spatial electron configuration of the atom.

In simple chemistry, as per valence bond theory, the carbon atom must follow the 4-hydrogen rule, which states that the maximum number of atoms available to bond with carbon is equal to the number of electrons that are attracted into the outer shell of carbon. In terms of shells, carbon consists of an incomplete outer shell, which comprises 4 electrons, and thus has 4 electrons available for covalent or dative bonding.

Hydrocarbons are hydrophobic like lipids.

Some hydrocarbons also are abundant in the solar system. Lakes of liquid methane and ethane have been found on Titan, Saturn's largest moon, confirmed by the Cassini-Huygens Mission. Hydrocarbons are also abundant in nebulae forming polycyclic aromatic hydrocarbon (PAH) compounds.

## Simple Hydrocarbons and their Variations

| Number of carbon atoms | Alkane (single bond) | Alkene (double bond) | Alkyne (triple bond) | Cycloalkane | Alkadiene |
|---|---|---|---|---|---|
| 1 | Methane | – | – | – | – |
| 2 | Ethane | Ethene (ethylene) | Ethyne (acetylene) | – | – |
| 3 | Propane | Propene (propylene) | Propyne (methylacetylene) | Cyclopropane | Propadiene (allene) |
| 4 | Butane | Butene (butylene) | Butyne | Cyclobutane | Butadiene |
| 5 | Pentane | Pentene | Pentyne | Cyclopentane | Pentadiene (piperylene) |
| 6 | Hexane | Hexene | Hexyne | Cyclohexane | Hexadiene |
| 7 | Heptane | Heptene | Heptyne | Cycloheptane | Heptadiene |
| 8 | Octane | Octene | Octyne | Cyclooctane | Octadiene |
| 9 | Nonane | Nonene | Nonyne | Cyclononane | Nonadiene |
| 10 | Decane | Decene | Decyne | Cyclodecane | Decadiene |
| 11 | Undecane | Undecene | Undecyne | Cycloundecane | Undecadiene |
| 12 | Dodecane | Dodecene | Dodecyne | Cyclododecane | Dodecadiene |

## Usage

Oil refineries are one way hydrocarbons are processed for use. Crude oil is processed in several stages to form desired hydrocarbons, used as fuel and in other products.

Hydrocarbons are a primary energy source for current civilizations. The predominant use of hydrocarbons is as a combustible fuel source. In their solid form, hydrocarbons take the form of asphalt (bitumen).

Tank wagon 33 80 7920 362-0 with hydrocarbon gas.

Mixtures of volatile hydrocarbons are now used in preference to the chlorofluorocarbons as a propellant for aerosol sprays, due to chlorofluorocarbons' impact on the ozone layer.

Methane ($CH_4$) and ethane ($C_2H_6$) are gaseous at ambient temperatures and cannot be readily liquefied by pressure alone. Propane ($C_3H_8$) is however easily liquefied, and exists in 'propane bottles' mostly as a liquid. Butane ($C_4H_{10}$) is so easily liquefied that it provides a safe, volatile fuel for small pocket lighters. Pentane ($C_5H_{12}$) is a colorless liquid at room temperature, commonly used in chemistry and industry as a powerful nearly odorless solvent of waxes and high molecular weight organic compounds, including greases. Hexane ($C_6H_{14}$) is also a widely used non-polar, non-aromatic solvent, as well as a significant fraction of common gasoline. The $C^6$ through $C^{10}$ alkanes, alkenes and isomeric cycloalkanes are the top components of gasoline, naphtha, jet fuel and specialized industrial solvent mixtures. With the progressive addition of carbon units, the simple non-ring structured hydrocarbons have higher viscosities, lubricating indices, boiling points, solidification temperatures, and deeper color. At the opposite extreme from methane lie the heavy tars that remain as the *lowest fraction* in a crude oil refining retort. They are collected and widely utilized as roofing compounds, pavement composition, wood preservatives (the creosote series) and as extremely high viscosity shear-resisting liquids.

Hydrocarbon use is also prevalent in nature. Some eusocial arthropods, such as the Brazilian stingless bee, *Schwarziana quadripunctata*, use unique hydrocarbon "scents" in order to determine kin from non-kin. The chemical hydrocarbon composition varies between age, sex, nest location, and hierarchal position.

## Reactions

There are three main types of reactions:

- Substitution reaction
- Addition reaction
- Combustion

## Substitution Reactions

Substitution reactions only occur in saturated hydrocarbons (single carbon–carbon bonds). In this reaction, an alkane reacts with a chlorine molecule. One of the chlorine atoms displaces a hydrogen atom. This forms hydrochloric acid as well as the hydrocarbon with one chlorine atom.

$$CH_4 + Cl_2 \rightarrow CH_3Cl + HCl$$

$$CH_3Cl + Cl_2 \rightarrow CH_2Cl_2 + HCl$$

all the way to $CCl_4$ (carbon tetrachloride).

$$C_2H_6 + Cl_2 \rightarrow C_2H_5Cl + HCl$$

$$C_2H_4Cl_2 + Cl_2 \rightarrow C_2H_3Cl_3 + HCl$$

all the way to $C_2Cl_6$ (hexachloroethane).

## Addition Reactions

Addition reactions involve alkenes and alkynes. In this reaction a halogen molecule breaks the double or triple bond in the hydrocarbon and forms a bond.

## Combustion

Hydrocarbons are currently the main source of the world's electric energy and heat sources (such as home heating) because of the energy produced when burnt. Often this energy is used directly as heat such as in home heaters, which use either petroleum or natural gas. The hydrocarbon is burnt and the heat is used to heat water, which is then circulated. A similar principle is used to create electrical energy in power plants.

Common properties of hydrocarbons are the facts that they produce steam, carbon dioxide and heat during combustion and that oxygen is required for combustion to take place. The simplest hydrocarbon, methane, burns as follows:

$$CH_4 + 2\,O_2 \rightarrow 2\,H_2O + CO_2 + \text{energy}$$

In inadequate supply of air, carbon monoxide gas and water vapour are formed:

$$2\,CH_4 + 3\,O_2 \rightarrow 2\,CO + 4\,H_2O$$

Another example is the combustion of propane:

$$C_3H_8 + 5\,O_2 \rightarrow 4\,H_2O + 3\,CO_2 + \text{energy}$$

And finally, for any linear alkane of n carbon atoms,

$$C_nH_{2n+2} + \frac{3n+1}{2}O_2 \rightarrow (n+1)H_2O + nCO_2 + \text{energy.}$$

Burning of hydrocarbons is an example of an exothermic chemical reaction.

Hydrocarbons can also be burned with elemental fluorine, resulting in carbon tetrafluoride and hydrogen fluoride products.

### Petroleum

Natural oil spring.

Extracted hydrocarbons in a liquid form are referred to as petroleum (literally "rock oil" – petrol for short) or mineral oil, whereas hydrocarbons in a gaseous form are referred to as natural gas. Petroleum and natural gas are found in the Earth's subsurface with the tools of petroleum geology and are a significant source of fuel and raw materials for the production of organic chemicals.

The extraction of liquid hydrocarbon fuel from sedimentary basins is integral to modern energy development. Hydrocarbons are mined from oil sands and oil shale, and potentially extracted from sedimentary methane hydrates. These reserves require distillation and upgrading to produce synthetic crude and petroleum.

Oil reserves in sedimentary rocks are the source of hydrocarbons for the energy, transport and petrochemical industries.

Economically important hydrocarbons include fossil fuels such as coal, petroleum and natural gas, and its derivatives such as plastics, paraffin, waxes, solvents and oils. Hydrocarbons – along with $NO_x$ and sunlight – contribute to the formation of tropospheric ozone and greenhouse gases.

### Bioremediation

Bacteria in the gabbroic layer of the ocean's crust can degrade hydrocarbons; but the extreme environment makes research difficult. Other bacteria such as *Lutibacterium anuloederans* can also degrade hydrocarbons. Mycoremediation or breaking down of hydrocarbon by mycelium and mushrooms is possible.

## Aliphatic Compound

An aliphatic compound or aliphatic hydrocarbon is an organic compound containing hydrogen and carbon atoms that are usually linked together in chains that are straight. Sometimes the chains are also in branched trains or in the form of non-aromatic structures. Notably, apart from hydrogen some of the other elements like oxygen, nitrogen, chlorine, and sulfur may be bound to the carbon atoms in the chain.

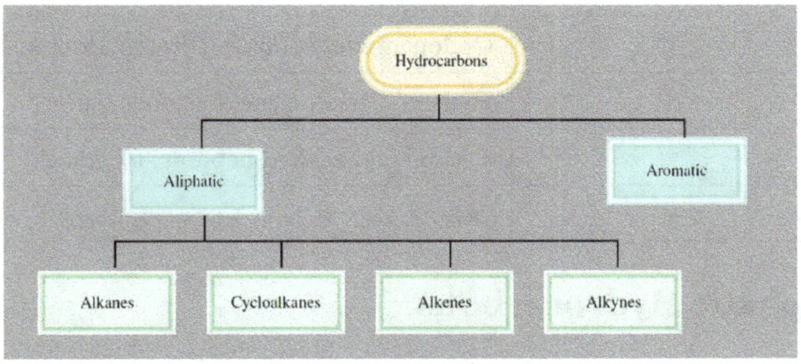


## Saturated and Unsaturated Hydrocarbons

Aliphatic compounds may be saturated or unsaturated. Saturated hydrocarbon contains mainly of alkanes which are open chain hydrocarbons containing carbon-carbon single bond. Most of the time the bond exists in the form of a covalent bond. These compounds are inert in nature and do not readily react with acid, bases or other reagents. Unsaturated hydrocarbons are alkenes and alkynes which have one carbon-carbon double bond and one carbon-carbon triple bond respectively. Unsaturated hydrocarbons are more reactive than saturated hydrocarbons and they usually fewer hydrogen atoms can be seen in bond with carbon atoms.

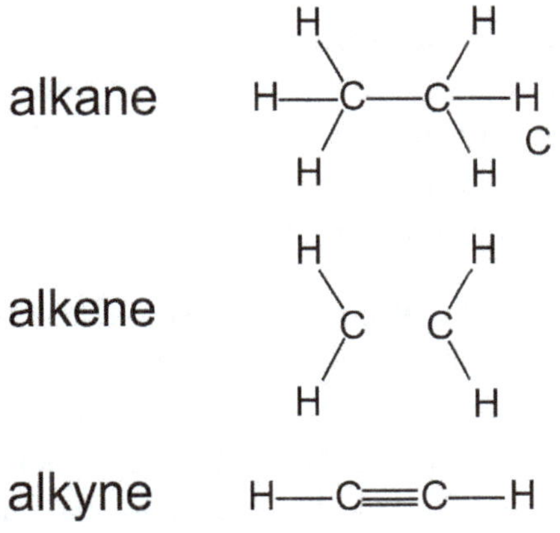


## Properties of Aliphatic Hydrocarbons

- Most aliphatic hydrocarbons are flammable. These compounds are used as fuels.
- Aliphatic compounds can be cyclic or acyclic meaning they can contain close chains or rings of carbon atoms in their molecule.

Examples And List of Aliphatic Hydrocarbons

| Number of Carbons | Aliphatic Hydrocarbons |
|---|---|
| 1 | Methane |
| 2 | Ethane, Ethene, Ethyne |

| | |
|---|---|
| 3 | Propane, Propene, Propyne, Cyclopropane |
| 4 | Methylpropane, Butane, Cyclobutene |
| 5 | Cyclopentene, Pentane, Dimethylpropane, |
| 6 | Hexane, Cyclohexene, Cyclohexane |
| 7 | Cycloheptane, Cycloheptene, Heptane |
| 8 | Octane, Cyclooctane, Cyclooctene |

### Extraction of Aliphatic Hydrocarbons

Aliphatic compounds can be extracted by the process known as Pressurized Fluid Extraction or PFE where organic and aqueous extraction solvents are used. Water which is converted to hot steam can also be used to extract aliphatic hydrocarbons mostly from solid and semi-solid environmental samples.

## Aromatic Compound

Aromatic compound is any of a large class of unsaturated chemical compounds characterized by one or more planar rings of atoms joined by covalent bonds of two different kinds. The unique stability of these compounds is referred to as aromaticity. Although the term aromatic originally concerned odour, today its use in chemistry is restricted to compounds that have particular electronic, structural, or chemical properties. Aromaticity results from particular bonding arrangements that cause certain $\pi$ (pi) electrons within a molecule to be strongly held. Aromaticity is often reflected in smaller than expected heats of combustion and hydrogenation and is associated with low reactivity.

Benzene ($C_6H_6$) is the best-known aromatic compound and the parent to which numerous other aromatic compounds are related. The six carbons of benzene are joined in a ring, having the planar geometry of a regular hexagon in which all of the C—C bond distances are equal. The six $\pi$ electrons circulate in a region above and below the plane of the ring, each electron being shared by all six carbons, which maximizes the force of attraction between the nuclei (positive) and the electrons (negative). Equally important is the number of $\pi$ electrons, which, according to molecular orbital theory, must be equal to $4n + 2$, in which $n = 1, 2, 3$, etc. For benzene with six $\pi$ electrons, $n = 1$.

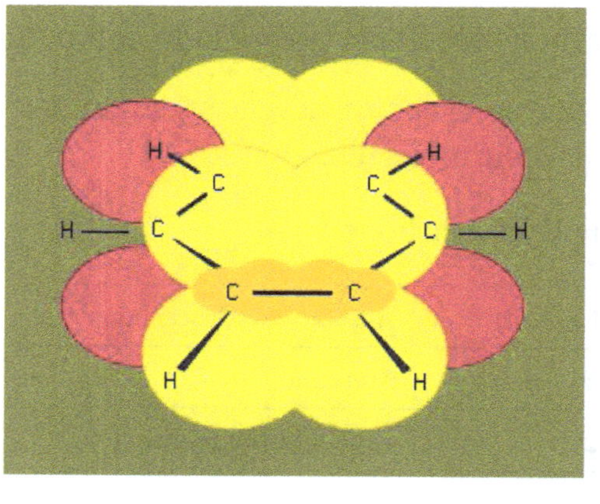

Benzene is the smallest of the organic aromatic hydrocarbons. It contains sigma bonds (represented by lines) and regions of high-pi electron density, formed by the overlapping of *p* orbitals (represented by the dark yellow shaded area) of adjacent carbon atoms, which give benzene its characteristic planar structure.

The largest group of aromatic compounds are those in which one or more of the hydrogens of benzene are replaced by some other atom or group, as in toluene ($C_6H_5CH_3$) and benzoic acid ($C_6H_5CO_2H$). Polycyclic aromatic compounds are assemblies of benzene rings that share a common side—for example, naphthalene ($C_{10}H_8$). Heterocyclic aromatic compounds contain at least one atom other than carbon within the ring. Examples include pyridine ($C_5H_5N$), in which one nitrogen (N) replaces one CH group, and purine ($C_5H_4N_4$), in which two nitrogens replace two CH groups. Heterocyclic aromatic compounds, such as furan ($C_4H_4O$), thiophene ($C_4H_4S$), and pyrrole ($C_4H_4NH$), contain five-membered rings in which oxygen (O), sulfur (S), and NH, respectively, replace an HC=CH unit.

# Phosgene

Phosgene is the organic chemical compound with the formula $COCl_2$. It is a colorless gas; in low concentrations, its odor resembles freshly cut hay or grass. Phosgene is a valued industrial building block, especially for the production of urethanes and polycarbonate plastics. However, it is very poisonous and was used as a chemical weapon during World War I where it was responsible for 85,000 deaths. In addition to its industrial production, small amounts occur from the breakdown and the combustion of organochlorine compounds.

## Structure and Basic Properties

Phosgene is a planar molecule as predicted by VSEPR theory. The C=O distance is 1.18 Å, the C–Cl distance is 1.74 Å and the Cl–C–Cl angle is 111.8°. It is one of the simplest acyl chlorides, being formally derived from carbonic acid.

## Production

Industrially, phosgene is produced by passing purified carbon monoxide and chlorine gas through a bed of porous activated carbon, which serves as a catalyst:

$$CO + Cl_2 \rightarrow COCl_2 \ (\Delta H_{rxn} = -107.6 \text{ kJ/mol})$$

The reaction is exothermic, therefore the reactor must be cooled. Typically, the reaction is conducted between 50 and 150 °C. Above 200 °C, phosgene reverts to carbon monoxide and chlorine, $K_{eq}$(300 K) = 0.05. World production of this compound was estimated to be 2.74 million tonnes in 1989.

Because of safety issues, phosgene is often produced and consumed within the same plant, and extraordinary measures are made to contain it. It is listed on Schedule 3 of the Chemical Weapons Convention: All production sites manufacturing more than 30 tonnes per year must be declared to the OPCW. Although less dangerous than many other chemical weapons such as sarin, phosgene is still regarded as a viable chemical warfare agent because it is so easy to manufacture when compared to the production requirements of more technically advanced chemical weapons such as the first-generation nerve agent tabun.

## Inadvertent Generation

Upon ultraviolet (UV) radiation in the presence of oxygen, chloroform slowly converts into phosgene by a radical reaction. To suppress this photodegradation, chloroform is often stored in brown-tinted glass containers and with a small percentage of ethanol added. Chlorinated solvents used to remove oil from metals, such as automotive brake cleaners, are converted to phosgene by the UV rays of arc welding processes.

Phosgene may also be produced during testing for leaks of older-style refrigerant gases. Chloromethanes (R12, R22 and others) were formerly leak-tested *in situ* by employing a small gas torch (propane, butane, or propylene gas) with a sniffer tube and a copper reaction plate in the flame nozzle of the torch. If any refrigerant gas was leaking from a pipe or joint, the gas would be sucked into the flame through the sniffer tube and would cause a colour change of the gas flame to a bright greenish blue. In the process, phosgene gas would be created due to the thermal reaction. No valid statistics are available, but anecdotal reports suggest that numerous refrigeration technicians suffered the effects of phosgene poisoning due to their ignorance of the toxicity of phosgene, produced during such leak testing. Electronic sensing of refrigerant gases phased out the use of flame testing for leaks in the 1980s. Similarly, phosgene poisoning is a possibility for people fighting fires that occur in the vicinity of refrigerant leaks from air-conditioning systems or refrigeration equipment, smoking in the vicinity of a freon refrigerant leak, or fighting fires using halon or halotron.

Phosgene can be released during building fires. In one instance, a deputy fire chief died ten days after inhaling fumes that wafted down outside a burning restaurant. After a two-day hospitalization he had appeared to recover, but ultimately suffered cardiac arrest at home following tracheobronchial inflammation, alveolar hemorrhage, and pulmonary edema. The phosgene was produced by decomposing Freon 22 after flames ducted up from a grease fire heated an air-conditioning unit on the roof and ruptured a hose.

## Uses

The great majority of phosgene is used in the production of isocyanates, the most important being toluene diisocyanate (TDI) and methylene diphenyl diisocyanate (MDI). These two isocyanates are precursors to polyurethanes.

## Synthesis of Carbonates

Significant amounts are also used in the production of polycarbonates by its reaction with bisphenol A. Polycarbonates are an important class of engineering thermoplastic found, for example, in lenses in eyeglasses. Diols react with phosgene to give either linear or cyclic carbonates (R = H, alkyl, aryl):

$$HOCR_2{-}X{-}CR_2OH + COCl_2 \rightarrow {}^1\!/_n\,[OCR_2{-}X{-}CR_2OC(O){-}]_n + 2\ HCl$$

### Synthesis of Isocyanates

The synthesis of isocyanates from amines illustrates the electrophilic character of this reagent and its use in introducing the equivalent of "$CO^{2+}$":

$$RNH_2 + COCl_2 \rightarrow RN{=}C{=}O + 2\ HCl\ (R = \text{alkyl, aryl})$$

Such reactions are conducted in the presence of a base such as pyridine that absorbs the hydrogen chloride.

### Laboratory uses

In the research laboratory phosgene still finds limited use in organic synthesis. A variety of substitutes have been developed, notably trichloromethyl chloroformate ("diphosgene"), a liquid at room temperature, and bis(trichloromethyl) carbonate ("triphosgene"), a crystalline substance. Aside from the above reactions that are widely practiced industrially, phosgene is also used to produce acyl chlorides and carbon dioxide from carboxylic acids:

$$RCO_2H + COCl_2 \rightarrow RC(O)Cl + HCl + CO_2$$

Such acid chlorides react with amines and alcohols to give, respectively, amides and esters, which are commonly used intermediates. Thionyl chloride is more commonly and more safely employed for this application. A specific application for phosgene is the production of chloroformic esters:

$$ROH + COCl_2 \rightarrow ROC(O)Cl + HCl$$

Phosgene is stored in metal cylinders. The outlet is always standard, a tapered thread that is known as CGA 160.

### Other Chemistry

Although it is somewhat hydrophobic, phosgene reacts with water to release hydrogen chloride and carbon dioxide:

$$COCl_2 + H_2O \rightarrow CO_2 + 2\ HCl$$

Analogously, with ammonia, one obtains urea:

$$COCl_2 + 4\ NH_3 \rightarrow CO(NH_2)_2 + 2\ NH_4Cl$$

Halide exchange with nitrogen trifluoride and aluminium tribromide gives COF2 and COBr2, respectively.

### Chemical Warfare

The collapse of international conventions against chemical weapons led to the widespread use of chlorine gas in World War I, but its lethal concentration of 0.1% was visible as a green cloud in the air, allowing troops to take readily available countermeasures. Phosgene, colorless with a more subtle "moldy hay" odor, was introduced by a group of French chemists led by Victor Grignard and first used by the French in 1915. It was also used in a mixture with an equal volume of chlorine, with the chlorine helping to spread the denser phosgene. Phosgene was more potent than chlorine, though some of the symptoms of exposure took 24 hours or more to manifest, meaning the victims were initially still capable of putting up a fight.

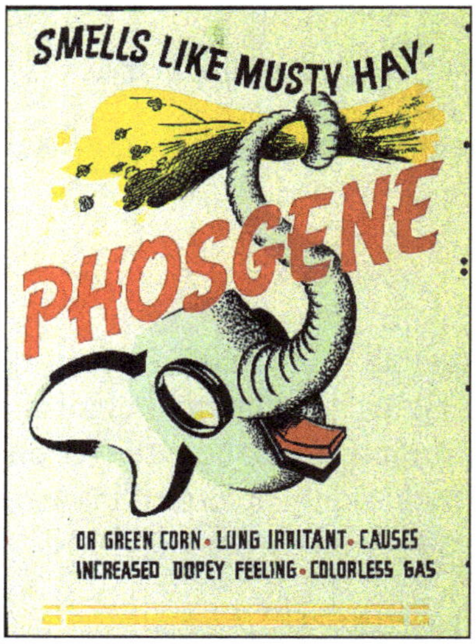


US Army phosgene identification poster from World War II.

Following the extensive use of phosgene gas in combat during World War I, it was stockpiled by various countries as part of their secret chemical weapons programs.

In May 1924, eleven tons of phosgene escaped from a war surplus store in central Hamburg. Three hundred people were poisoned, of whom 10 died.

Phosgene was then only frequently used by the Imperial Japanese Army against the Chinese during the Second Sino-Japanese War. Gas weapons, such as phosgene, were produced by Unit 731 and authorized by specific orders given by Hirohito (Emperor Showa) himself, transmitted by the chief of staff of the army. For example, the Emperor authorized the use of toxic gas on 375 separate occasions during the Battle of Wuhan from August to October 1938.

## Formaldehyde

Formaldehyde (systematic name methanal) is a naturally occurring organic compound with the formula $CH_2O$ (H–CHO). It is the simplest of the aldehydes (R–CHO). The common name of this substance comes from its similarity and relation to formic acid.

Formaldehyde is an important precursor to many other materials and chemical compounds. In 1996, the installed capacity for the production of formaldehyde was estimated at 8.7 million tons per year. It is mainly used in the production of industrial resins, e.g., for particle board and coatings.

In view of its widespread use, toxicity, and volatility, formaldehyde poses a significant danger to human health. In 2011, the US National Toxicology Program described formaldehyde as "known to be a human carcinogen".

## Forms of Formaldehyde

Formaldehyde is more complicated than many simple carbon compounds in that it adopts several different forms. As a gas, formaldehyde is colorless and has a characteristic pungent, irritating odor. Upon condensation, the gas converts to various other forms of formaldehyde (with different chemical formulas) that are of more practical value. One important derivative is the cyclic trimer metaformaldehyde (1,3,5-trioxane) with the formula $(CH_2O)_3$. There is also a linear polymer called paraformaldehyde. These compounds have similar chemical properties and are often used interchangeably.

When dissolved in water, formaldehyde also forms a hydrate, methanediol, with the formula $H_2C(OH)_2$. This compound also exists in equilibrium with various oligomers (short polymers), depending on the concentration and temperature. A saturated water solution, of about 40% formaldehyde by volume or 37% by mass, is called “100% formalin”. A small amount of stabilizer, such as methanol, is usually added to suppress oxidation and polymerization. A typical commercial grade formalin may contain 10–12% methanol in addition to various metallic impurities. The name was genericised long ago from an old trade name, “Formalin”.

## Occurrence

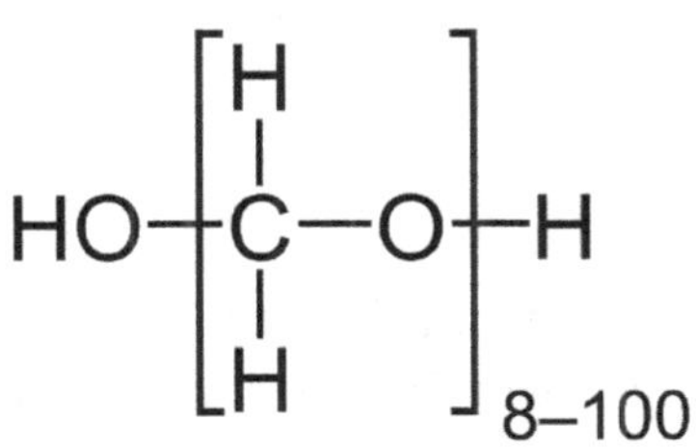

Paraformaldehyde is a common form of formaldehyde for industrial applications.

Processes in the upper atmosphere contribute up to 90% of the total formaldehyde in the environment. Formaldehyde is an intermediate in the oxidation (or combustion) of methane, as well as of other carbon compounds, e.g. in forest fires, automobile exhaust, and tobacco smoke. When produced in the atmosphere by the action of sunlight and oxygen on atmospheric methane and other hydrocarbons, it becomes part of smog. Formaldehyde has also been detected in outer space.

Formaldehyde and its adducts are ubiquitous in living organisms. It is formed in the metabolism of endogenous amino acids and is found in the bloodstream of humans and other primates at concentrations of approximately 0.1 millimolar. Experiments in which animals are exposed to an atmosphere containing isotopically labeled formaldehyde have demonstrated that even in deliberately exposed animals, the majority of formaldehyde-DNA adducts found in non-respiratory tissues are derived from endogenously produced formaldehyde.

Formaldehyde does not accumulate in the environment, because it is broken down within a few hours by sunlight or by bacteria present in soil or water. Humans metabolize formaldehyde quickly, converting it to formic acid, so it does not accumulate in the body.

## Interstellar Formaldehyde

Formaldehyde appears to be a useful probe in astrochemistry due to prominence of the $1_{10}\leftarrow1_{11}$ and $2_{11}\leftarrow2_{12}$ *K*-doublet transitions. It was the first polyatomic organic molecule detected in the interstellar medium. Since its initial detection in 1969, it has been observed in many regions of the galaxy. Because of the widespread interest in interstellar formaldehyde, it has been extensively studied, yielding new extragalactic sources. A proposed mechanism for the formation is the hydrogenation of CO ice:

$$H + CO \rightarrow HCO$$

$$HCO + H \rightarrow CH_2O$$

HCN, HNC, $H_2CO$, and dust have also been observed inside the comae of comets C/2012 F6 (Lemmon) and C/2012 S1 (ISON).

## Synthesis and Industrial Production

### Industry

Formaldehyde is produced industrially by the catalytic oxidation of methanol. The most common catalysts are silver metal or a mixture of an iron and molybdenum or vanadium oxides. In the commonly used formox process, methanol and oxygen react at ca. 250–400 °C in presence of iron oxide in combination with molybdenum and/or vanadium to produce formaldehyde according to the chemical equation:

$$2\,CH_3OH + O_2 \rightarrow 2\,CH_2O + 2\,H_2O$$

The silver-based catalyst usually operates at a higher temperature, about 650 °C. Two chemical reactions on it simultaneously produce formaldehyde: that shown above and the dehydrogenation reaction:

$$CH_3OH \rightarrow CH_2O + H_2$$

In principle, formaldehyde could be generated by oxidation of methane, but this route is not industrially viable because the methanol is more easily oxidized than methane.

## Organic Chemistry

Formaldehyde is a building block in the synthesis of many other compounds of specialised and industrial significance. It exhibits most of the chemical properties of other aldehydes but is more reactive.

### Self-condensation and Hydration

Formaldehyde, unlike most aldehydes, oligomerizes spontaneously. The trimer is trioxane, and the polymer is called paraformaldehyde. Many cyclic oligomers have been isolated. Similarly, formaldehyde hydrates to give the geminal diol methanediol, which condenses further to form oligomers $HO(CH_2O)_nH$. Monomeric $CH_2O$ is rarely encountered.

### Oxidation

It is readily oxidized by atmospheric oxygen into formic acid. For this reason, commercial formaldehyde is typically contaminated with formic acid.

## Hydroxymethylation and Chloromethylation

Formaldehyde is a good electrophile. With good nucleophiles such as thiols, amines, and even amides, no acid catalyst is required. The resulting hydroxymethyl derivatives typically react further. Thus amines give hexahydro-1,3,5-triazines. Similarly, when combined with hydrogen sulfide, it forms trithiane.

$$3CH_2O + 3H_2S \rightarrow (CH_2S)_3 + 3H_2O$$

In the presence of acids, it participates in electrophilic aromatic substitution reactions with aromatic compounds resulting in hydroxymethylated derivatives:

$$ArH + CH_2O \rightarrow ArCH_2OH$$

When conducted in the presence of hydrogen chloride, the product is the chloromethyl compound, as described in the Blanc chloromethylation. If the arene is electron-rich, as in phenols, elaborate condensations ensue. With 4-substituted phenols one obtains calixarenes. Phenol results in polymers.

## Base Reactions

Cannizzaro reaction in the presence of basic catalysts to produce formic acid and methanol.

## Uses

## Industrial Applications

Formaldehyde is a common precursor to more complex compounds and materials. In approximate order of decreasing consumption, products generated from formaldehyde include urea formaldehyde resin, melamine resin, phenol formaldehyde resin, polyoxymethylene plastics, 1,4-butanediol, and methylene diphenyl diisocyanate. The textile industry uses formaldehyde-based resins as finishers to make fabrics crease-resistant. Formaldehyde-based materials are key to the manufacture of automobiles, and used to make components for the transmission, electrical system, engine block, door panels, axles and brake shoes. The value of sales of formaldehyde and derivative products was over $145 billion in 2003, about 1.2% of the gross domestic product (GDP) of the United States and Canada. Including indirect employment, over 4 million people work in the formaldehyde industry across approximately 11,900 plants in the U.S. and Canada.

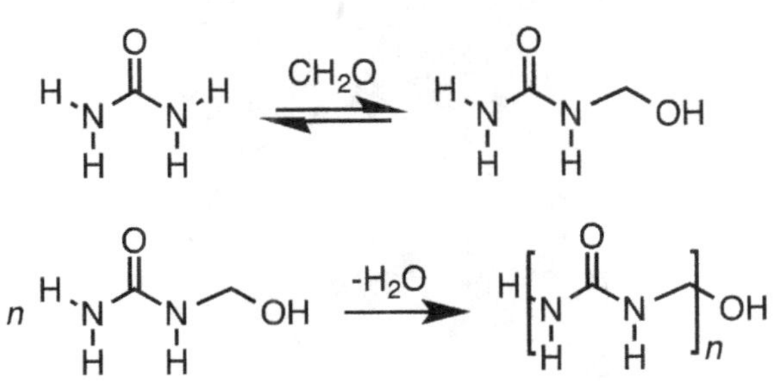

Two steps in formation of urea-formaldehyde resin, which is widely used in the production of particle board.

When treated with phenol, urea, or melamine, formaldehyde produces, respectively, hard thermoset phenol formaldehyde resin, urea formaldehyde resin, and melamine resin. These polymers are common permanent adhesives used in plywood and carpeting. It is used as the wet-strength resin added to sanitary paper products such as (listed in increasing concentrations injected into the paper machine headstock chest) facial tissue, table napkins, and roll towels. They are also foamed to make insulation, or cast into moulded products. Production of formaldehyde resins accounts for more than half of formaldehyde consumption.

Formaldehyde is also a precursor to polyfunctional alcohols such as pentaerythritol, which is used to make paints and explosives. Other formaldehyde derivatives include methylene diphenyl diisocyanate, an important component in polyurethane paints and foams, and hexamine, which is used in phenol-formaldehyde resins as well as the explosive RDX.

Condensation with acetaldehyde affords pentaerythritol, a chemical necessary in synthesizing PETN, a high explosive. Condensation with phenols gives phenol-formaldehyde resins.

## Niche uses

### Disinfectant and Biocide

An aqueous solution of formaldehyde can be useful as a disinfectant as it kills most bacteria and fungi (including their spores). It is used to kill produced vaccines. Formaldehyde releasers are used as biocides in personal care products such as cosmetics. Although present at levels not normally considered harmful, they are known to cause allergic contact dermatitis in certain sensitised individuals.

Aquarists use formaldehyde as a treatment for the parasites *Ichthyophthirius multifiliis* and *Cryptocaryon irritans*.

Formaldehyde is also approved for use in the manufacture of animal feeds in the US. It is an antimicrobial agent used to maintain complete animal feeds or feed ingredients Salmonella negative for up to 21 days.

### Tissue Fixative and Embalming Agent

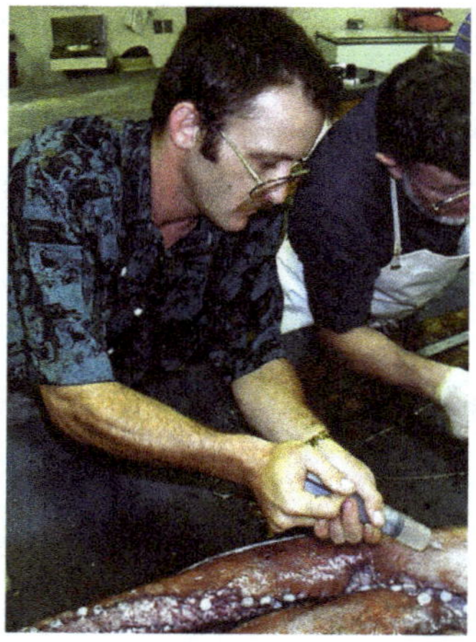

Injecting a giant squid specimen with formalin for preservation.

Formaldehyde preserves or fixes tissue or cells. The process involves cross-linking of primary amino groups. The European Union has banned the use of formaldehyde due to its carcinogenic properties as a biocide (including embalming) under the Biocidal Products Directive (98/8/EC). Countries with a strong tradition of embalming corpses, such as Ireland and other colder-weather countries, have raised concerns. Despite reports to the contrary, no decision on the inclusion of formaldehyde on Annex I of the Biocidal Products Directive for product-type 22 (embalming and taxidermist fluids) had been made as of September 2009.

Formaldehyde-based crosslinking is exploited in ChIP-on-chip or ChIP-sequencing genomics experiments, where DNA-binding proteins are cross-linked to their cognate binding sites on the chromosome and analyzed to determine what genes are regulated by the proteins. Formaldehyde is also used as a denaturing agent in RNA gel electrophoresis, preventing RNA from forming secondary structures. A solution of 4% formaldehyde fixes pathology tissue specimens at about one mm per hour at room temperature.

### Drug Testing

Formaldehyde and an 18 M (concentrated) sulfuric acid makes Marquis reagent—which can identify alkaloids and other compounds.

### Photography

In photography, formaldehyde is used in low concentrations for process C-41 (color negative film) stabilizer in the final wash step, as well as in the process E-6 pre-bleach step, to make it unnecessary in the final wash.

### Astringent

Formaldehyde topical can be used as an antiperspirant to treat hyperhydrosis.

## Heterocyclic Compound

Heterocyclic compound is also called heterocycle, any of a major class of organic chemical compounds characterized by the fact that some or all of the atoms in their molecules are joined in rings containing at least one atom of an element other than carbon (C). The cyclic part of heterocyclic indicates that at least one ring structure is present in such a compound, while the prefix hetero- refers to the noncarbon atoms, or heteroatoms, in the ring. In their general structure, heterocyclic compounds resemble cyclic organic compounds that incorporate only carbon atoms in the rings—for example, cyclopropane (with a three-carbon-atom ring) or benzene (with a six-carbon-atom ring)—but the presence of the heteroatoms gives heterocyclic compounds physical and chemical properties that are often quite distinct from those of their all-carbon-ring analogs.

Heterocyclic compounds include many of the biochemical material essential to life. For example, nucleic acids, the chemical substances that carry the genetic information controlling inheritance, consist of long chains of heterocyclic units held together by other types of materials. Many naturally occurring pigments, vitamins, and antibiotics are heterocyclic compounds, as are most hallucinogens. Modern society is dependent on synthetic heterocycles for use as drugs, pesticides, dyes, and plastics.

## General Aspects of Heterocyclic Compounds

The most common heterocycles are those having five- or six-membered rings and containing heteroatoms of nitrogen (N), oxygen (O), or sulfur (S). The best known of the simple heterocyclic compounds are pyridine, pyrrole, furan, and thiophene. A molecule of pyridine contains a ring of six atoms—five carbon atoms and one nitrogen atom. Pyrrole, furan, and thiophene molecules each contain five-membered rings, composed of four atoms of carbon and one atom of nitrogen, oxygen, or sulfur, respectively.

Pyridine and pyrrole are both nitrogen heterocycles—their molecules contain nitrogen atoms along with carbon atoms in the rings. The molecules of many biological materials consist in part of pyridine and pyrrole rings, and such materials yield small amounts of pyridine and pyrrole upon strong heating. In fact, both of these substances were discovered in the 1850s in an oily mixture formed by strong heating of bones. Today, pyridine and pyrrole are prepared by synthetic reactions.Their chief commercial interest lies in their conversion to other substances, chiefly dyestuffs and drugs. Pyridine is used also as a solvent, a waterproofing agent, a rubber additive, an alcohol denaturant, and a dyeing adjunct.

Furan is an oxygen-containing heterocycle employed primarily for conversion to other substances (including pyrrole). Furfural, a close chemical relative of furan, is obtained from oat hulls and corncobs and is used in the production of intermediates for nylon. Thiophene, a sulfur heterocycle, resembles benzene in its chemical and physical properties. It is a frequent contaminant of the benzene obtained from natural sources and was first discovered during the purification of benzene. Like the other compounds, it is used primarily for conversion to other substances. Furan and thiophene were both discovered in the latter part of the 19th century.

In general, the physical and chemical properties of heterocyclic compounds are best understood by comparing them with ordinary organic compounds that do not contain heteroatoms.

### Comparison with Carbocyclic Compounds

The molecules of organic chemical compounds are built up from a framework or backbone of carbon atoms to which are attached hydrogen (H), oxygen, or other heteroatoms. Carbon atoms have the unique property of being able to link with one another to form chains of atoms. When the ends of the chains are joined together into a ring, cyclic compounds result; such substances often are referred to as carbocyclic or alicyclic compounds. Substitution of one or more of the ring carbon atoms in the molecules of a carbocyclic compound with a heteroatom gives a heterocyclic compound.

A typical carbocyclic compound is cyclopentane ($C_5H_{10}$), the molecular structure of which is indicated by the formula:

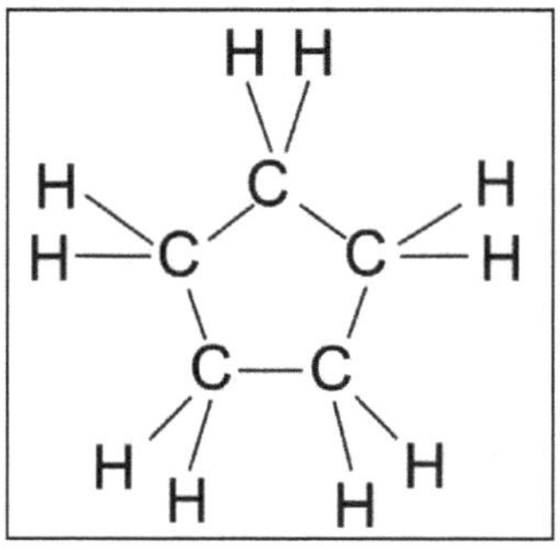

In which the chemical symbols represent atoms of the elements and the lines represent bonds between the atoms. For convenience such formulas are often written in simplified polygonal form, such as:

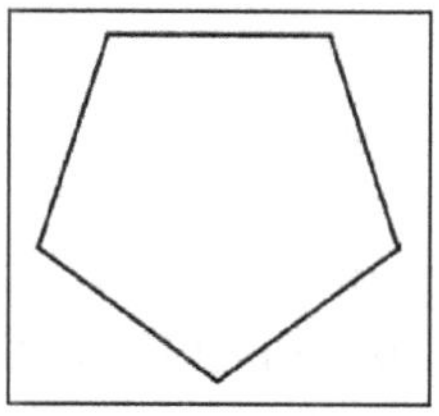

For cyclopentane, in which each corner of the polygon represents a carbon atom (it being understood that hydrogen atoms are joined to the carbon atoms as required).

When one of the carbon atoms of cyclopentane is replaced with an atom of nitrogen, the compound pyrrolidine, a chemical relative of pyrrole, is produced. The structural formula of pyrrolidine is written:

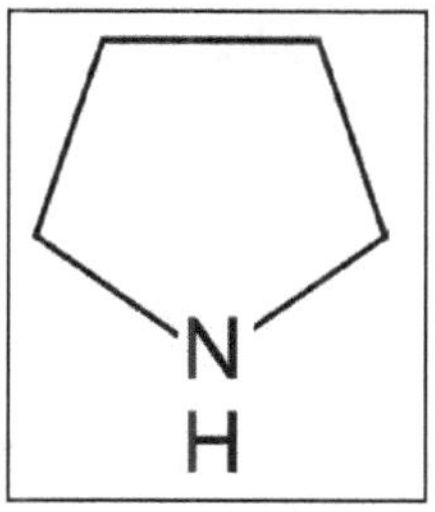


Other heterocyclic compounds can be envisioned similarly as derivatives of cyclopentane by substitution with other heteroatoms or of other carbocyclic compounds by substitution with nitrogen or other heteroatoms.

The simplest organic compounds are the hydrocarbons, compounds of carbon and hydrogen only. Hydrocarbons are classified as saturated if all four possible bonds of every carbon atom are joined singly to another carbon atom or to a hydrogen atom. They are classified as unsaturated if they contain a double or triple bond between any two of the carbon atoms, and they are classed as aromatic if they contain at least one ring, all atoms of which are joined by alternating double and single bonds. Nonaromatic unsaturated compounds are highly reactive—that is, they readily undergo additions of atoms or groups of atoms to the carbon atoms of their double bonds, giving each carbon four substituents. Aromatic compounds, though having double bonds, are extremely stable and do not readily undergo the addition reactions characteristic of other unsaturated compounds. The stability and unreactivity of, for instance, a six-membered aromatic ring are associated with the presence of three pairs of electrons, called pi ($\pi$) electrons, associated with the three double bonds of the ring. Together these electrons, constituting the so-called aromatic sextet, form an unusually stable structure associated with the aromatic ring as a whole rather than with the individual pairs of atoms.

Heterocycles too may be classified as saturated, unsaturated, or aromatic. Thus, as shown in the following structural formulas, pyrrolidine is a saturated heterocyclic compound containing no double bond; 4,5-dihydrofuran is an unsaturated heterocyclic compound; and pyridine is a typical heterocyclic aromatic, or heteroaromatic, substance. In the two structural formulas given for pyridine, the first shows the double bonds, whereas the second represents the aromatic sextet with a circle.

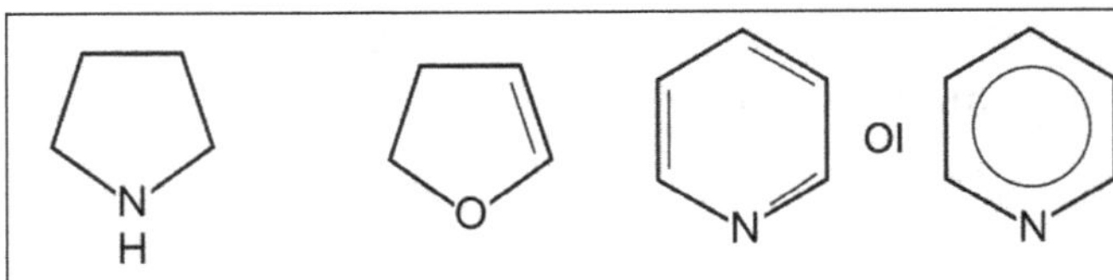


pyrrolidine 4,5-dihydrofuram pyridine

This classification relates the chemistry of heterocycles directly with that of carbocyclic derivatives, which are usually better known. In general, synthetic methods and physical and chemical properties of the saturated and the partly unsaturated heterocyclic compounds closely resemble those for their acyclic (noncyclic, or open-chain) analogs. Thus, pyrrolidine may be considered as a cyclic secondary amine and has much in common with the corresponding acyclic amine, diethylamine, which is represented by the formula:

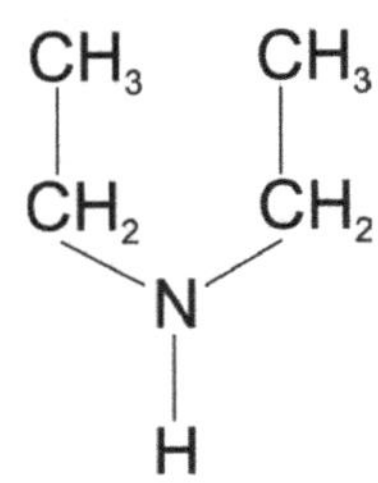


Similarly, 4,5-dihydrofuran mirrors many of the properties of the corresponding unsaturated ether, ethyl vinyl ether, which has the formula:

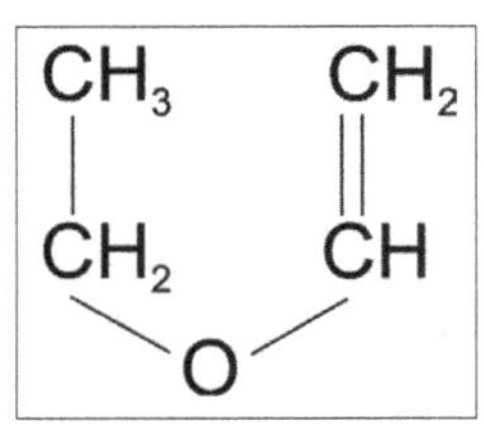


## Nomenclature of Heterocyclic Compounds

Naming heterocyclic compounds is complicated because of the existence of many common names in addition to the internationally agreed-upon systematic nomenclature.

The types of heteroatoms present in a ring are indicated by prefixes; in particular, oxa-, thia-, and aza- denote oxygen, sulfur, and nitrogen atoms, respectively. The numbers of heteroatoms of a particular kind are indicated by number prefixes joined to the heteroatom prefixes, as dioxa- and triaza-. The presence of different kinds of heteroatoms is indicated by combining the above prefixes, using the following order of preference: oxa- first, followed by thia- and then aza-.

In addition, partially saturated rings are indicated by the prefixes dihydro-, tetrahydro-, and so on, according to the number of "extra" hydrogen atoms bonded to the ring atoms. The positions of heteroatoms, extra hydrogen atoms, and substituents are indicated by Arabic numerals, for which the numbering starts at an oxygen atom, if one is present, or at a sulfur or nitrogen atom and continues in such a way that the heteroatoms are assigned the lowest possible numbers. Other things being equal, numbering starts at a nitrogen atom that carries a substituent rather than at a multiply bonded nitrogen. In compounds with maximum unsaturation, if the double bonds can be arranged in more than one way, their positions are defined by indicating the nitrogen or carbon atoms that are not multiply bonded and that consequently carry an extra hydrogen atom (or substituent), as follows: 1*H*-, 2*H*-, and so on.

## The Nature of Heteroaromaticity

Aromaticity denotes the significant stabilization of a ring compound by a system of alternating single and double bonds—called a cyclic conjugated system—in which six $\pi$ electrons generally participate. A nitrogen atom in a ring can carry a positive or a negative charge, or it can be in the neutral form. An oxygen or sulfur atom in a ring can either be in the neutral form or carry a positive charge. A fundamental distinction is usually made between (1) those heteroatoms that participate in a cyclic conjugated system by means of a lone, or unshared, pair of electrons that are in an orbital perpendicular to the plane of the ring and (2) those heteroatoms that do so because they are connected to another atom by means of a double bond.

An example of an atom of the first type is the nitrogen atom in pyrrole, which is linked by single covalent bonds to two carbon atoms and one hydrogen atom. Nitrogen has an outermost shell of five electrons, three of which can enter into three covalent bonds with other atoms. After the bonds are formed, as in the case of pyrrole, there remains an unshared electron pair that can engage in cyclic conjugation. The aromatic sextet in pyrrole is made up of two electrons from each of the two carbon-carbon double bonds and the two electrons that compose the unshared electron pair of the nitrogen atom. As a consequence, there tends to be a net flow of electron density from the nitrogen atom to the carbon atoms as the nitrogen's electrons are drawn into the aromatic sextet.

Alternatively, the pyrrole molecule may be described as a resonance hybrid—that is, a molecule whose true structure can only be approximated by two or more different forms, called resonance forms.

An example of a heteroatom of the second type is the nitrogen atom in pyridine, which is linked by covalent bonds to only two carbon atoms. Pyridine also has a $\pi$-electron sextet, but the nitrogen atom contributes only one electron to it, one additional electron being contributed by each of the five carbon atoms in the ring. In particular, the unshared electron pair of the nitrogen atom is not involved. Moreover, because nitrogen's attraction for electrons (its electronegativity) is greater than that of carbon, electrons tend to move toward the nitrogen atom rather than away from it, as in pyrrole.

Quite generally, heteroatoms may be referred to as pyrrolelike or pyridine-like, depending on whether they fall into the first or second class. The pyrrolelike heteroatoms $-NR-$ (R being hydrogen or a hydrocarbon group), $-N^{-}-$, $-O-$, and $-S-$ tend to donate electrons into the $\pi$-electron system, whereas the pyridine-like heteroatoms $-N=$, $-N^{+}R=$, $-O^{+}=$, and $-S^{+}=$ tend to attract the $\pi$ electrons of a double bond.

In six-membered heteroaromatic rings, the heteroatoms (usually nitrogen) are pyridine-like—for example, the compounds pyrimidine, which contains two nitrogen atoms, and 1,2,4-triazine, which contains three nitrogen atoms.

Six-membered heteroaromatic compounds cannot normally contain pyrrolelike heteroatoms. Five-membered heteroaromatic rings, however, always contain one pyrrolelike nitrogen, oxygen, or sulfur atom, and they may also contain up to four pyridine-like heteroatoms, as in the compounds thiophene (with one sulfur atom), 1,2,4-oxadiazole (with one oxygen atom and two nitrogen atoms), and pentazole (with five nitrogen atoms).

The quantitative measurement of aromaticity—and even its precise definition—has challenged chemists since German chemist August Kekule formulated the ring structure for benzene in the mid-19th century. Various methods based on energetic, structural, and magnetic criteria have been widely used to measure the aromaticity of carbocyclic compounds. All of them, however, are difficult to apply quantitatively to heteroaromatic systems because of complications arising from the presence of heteroatoms.

Chemical reactivity can provide a certain qualitative insight into aromaticity. The reactivity of an aromatic compound is affected by the extra stability of the conjugated system that it contains; the extra stability in turn determines the tendency of the compound to react by substitution of hydrogen—i.e., replacement of a singly bonded hydrogen atom with another singly bonded atom or group—rather than by addition of one or more atoms to the molecule via the breaking of a double bond. In terms of reactivity, therefore, the degree of aromaticity is measured by the relative tendency toward substitution rather than addition. By this criterion, pyridine is more aromatic than furan, but it is difficult to say just how much more aromatic.

## Physical Properties of Heterocyclic Compounds

Physical properties are important as criteria for judging the purity of heterocycles just as for other organic compounds. Organic compounds generally show great regularity in their physical properties, and heterocycles are no exception.

The melting point was once a widely used criterion for purity, but it has been increasingly superseded by optical spectra, based on light absorption; mass spectra, based on relative masses of molecular fragments; and magnetic resonance spectra, based on nuclear properties. Nevertheless, knowledge of melting and boiling points is still helpful for judging the purity of a compound.

## Melting and Boiling Points

The boiling points of certain saturated heterocycles are compared with those of the corresponding cycloalkanes. The melting points or boiling points of common heteroaromatic compounds and their substituted derivatives are compared with those of benzene.

| Ring size | Number (and position) of heteroatoms | Type of heteroatom | | | Saturated cycloalkane |
|---|---|---|---|---|---|
| | | N (as NH) | O | S | |
| 3 | one | 56 | 11 | 55 | −33 |
| 4 | one | 63 | 48 | 94 | 13 |
| 5 | one | 87 | 65 | 121 | 49 |
| 6 | one | 106 | 88 | 141 | 80 |

| | | | | | |
|---|---|---|---|---|---|
| 6 | two (1,2) | 150 | 116 | 190* | 80 |
| 6 | two (1,3) | 150 | 106 | 207 | 80 |
| 6 | two (1,4) | 145 | 101 | 200 | 80 |
| 7 | one | 138 | 120 | 174 | 119 |

Replacement of a two-carbon unit (two carbon and two hydrogen atoms, molecular weight equal to 26) by a single sulfur atom (atomic weight 32) has little effect on the melting or boiling point. On the other hand, replacement of a two-carbon unit by an oxygen atom (atomic weight 16) lowers the boiling point by about 40 °C (72 °F), which is to be expected because of the decreased molecular weight of the furan compounds (lighter compounds being more volatile). Introduction of nitrogen atoms into the benzene ring is accompanied by less-regular changes. Replacement of a two-carbon unit by an imino (NH) group, or of a single carbon by a nitrogen atom, increases the boiling point. Furthermore, making these two changes simultaneously increases the boiling point even more, probably as the result of intermolecular association by hydrogen bonding (a weak form of attachment via certain types of hydrogen atoms; between the pyridine-like nitrogen atom and the imino group.

The effects of substituent groups in heteroaromatic rings show considerable regularity. Methyl ($CH_3$) and ethyl ($C_2H_5$) groups attached to ring carbon atoms usually increase the boiling point by about 20–30 °C (36–54 °F) and 50–60 °C (90–108 °F), respectively, whereas a similar attachment to a ring nitrogen atom (e.g., pyrrole → 1-methylpyrrole) significantly decreases the boiling point because of decreased ease of intermolecular association by hydrogen bonding (the active hydrogen having been replaced by a hydrocarbon group). Heterocyclic carboxylic acids and amides are all solids at room temperature. Carboxylic acids of heterocycles containing a ring nitrogen atom usually melt at higher temperatures than those containing ring oxygen or sulfur atoms, because of hydrogen bonding. Compounds containing both a ring nitrogen atom and a hydroxyl (OH) or amino ($NH_2$) group are usually relatively high-melting solids. Compounds containing chlorine (Cl) usually have boiling points similar to those of the corresponding ethyl-substituted compounds.

## Ultraviolet, Infrared, Nuclear Magnetic Resonance and Mass Spectra

Spectroscopic studies of heterocyclic compounds, like those of other organic compounds, have became of great importance as means of identification of unknown materials, as criteria for purity, and as probes for investigating the electronic structures of molecules, thereby explaining and helping to predict their reactions. The ultraviolet spectrum of an organic compound (the pattern of its light absorption in the ultraviolet region of the spectrum) is characteristic of the π-electron system of the molecule—i.e., of the arrangement of double bonds within the structure. The ultraviolet spectra of heteroaromatic compounds show general similarity to those of benzenoid compounds (compounds with one or more benzene rings), and the effects of substituents can usually be rationalized in a similar way.

The infrared spectrum of an organic compound, with its complexity of bands, provides an excellent "fingerprint" of the compound—far more characteristic than a melting point. It also can be used to identify certain common groups, such as carbonyl (C=O) and imino, as well as various heterocyclic ring systems.

Magnetic resonance spectra are indispensable today for studies in heterocyclic chemistry. Proton resonance spectra, the most common type, yield information regarding the number of hydrogen atoms in the molecule, their chemical environment, and their relative orientation in space. Mass spectra are used to determine not only the complete molecular formula of the compound but also the molecular structure from the way the molecule fragments.

## Synthesis and Modification of Heterocyclic Rings

The important methods for synthesizing heterocyclic compounds can be classified under five headings. Three are ways of forming new heterocyclic rings from precursors containing either no rings (acyclic precursors) or one fewer ring than the desired product; one is a way of obtaining a heterocyclic ring from another heterocyclic ring or from a carbocyclic ring; and one involves the modification of substituents on an existing heterocyclic ring.

In the formation of rings from acyclic precursors, the key step is frequently the formation of a carbon-heteroatom linkage (C−Z, in which Z represents an atom of nitrogen, oxygen, sulfur, or a more unusual element). The actual ring closure, or cyclization, however, may involve the formation of a carbon-carbon bond. In any case, ring formation reactions are divided into three general categories according to whether the cyclization reaction occurs primarily as a result of nucleophilic or electrophilic attack or by way of a cyclic transition state.

## Nucleophilic Ring Closure

To prepare compounds containing one heteroatom, an open-chain hydrocarbon derivative containing two halogen element atoms—specifically, chlorine, bromine (Br), or iodine (I)—either as halides (in which the halogen atoms are attached directly to the hydrocarbon chain) or as acyl halides (in which the halogen atoms belong to derivatives of carboxylic acids) is reacted with the dihydro form of the heteroatom ($ZH_2$, or an equivalent reagent) to give nonaromatic heterocycles.

Diketones also can react with dihydro Z compounds to give heterocycles. (A ketone is an organic compound that contains a carbonyl group, the carbon atom of which is linked to two other carbon atoms belonging to hydrocarbon groups. Diketones contain two such carbonyl groups.) Diketones with the carbonyl groups separated by two carbon atoms, for example, can be cyclized to form five-membered aromatic pyrroles, furans, and thiophenes. In the case of diketones whose carbonyl groups are separated by three carbons, six-membered rings may be formed.

In each of these reactions the heteroatom Z acts as a nucleophile—an atom or a molecule that seeks a positively charged centre, such as a partly unprotected atomic nucleus. The heteroatom attacks the positively charged carbon atom produced by electron withdrawal because of the presence of the halogen atom or of the oxygen atom.

Usually, such reactions proceed by means of intermediates in which only one of the two C−Z bonds has formed. In reactions involving halogens as halides, for instance, a compound such as $HZ-CH_2-(CH_2)n-CH_2Br$ may form first. This fact can be applied to heterocycle synthesis in that it is frequently possible to make such intermediate compounds by other routes; these intermediates then cyclize readily to form the desired ring. One procedure for pyridine synthesis, for example, involves a condensation reaction employing an intermediate with the carbon-nitrogen bond already formed.

Heterocycles containing two adjacent nitrogen atoms, two oxygen atoms, or adjacent nitrogen and oxygen atoms also may be prepared from precursors by the use of hydrazine ($N_2H_4$), hydroxylamine ($NH_2OH$), or hydrogen peroxide ($H_2O_2$) in place of the dihydro Z compound. Similarly, two adjacent heteroatoms can be introduced by employing one of the reagents in the reactions with diketones.

When a compound containing two nonadjacent heteroatoms is desired, appropriate components can be put together, as in the synthesis of a pyrimidine. Ring synthesis reactions in which the heteroatom acts as a nucleophile can also employ precursors containing a ring, resulting in two-ring compounds. These reactions involve the use of *ortho*-disubstituted benzenes. The formation of quinoline and quinazoline rings is an example of this reaction type.

## Electrophilic Ring Closure

Heterocyclic ring-forming reactions in which the heteroatom acts as an electrophile—an electron-seeking atom or molecule—are rare, because nitrogen, oxygen, and sulfur atoms are themselves electron-rich centres that act generally as nucleophiles. Nevertheless, electrophilic ring closure reactions are known in which a heterocyclic ring is formed by a reaction in which a carbon atom of the future ring acts as an electrophile. Usually such reactions involve ring closure onto an existing benzene ring (or other aromatic system), an electron-rich system that is generally subject to attack by electrophilic reagents. An example of ring closures of this type is the formation of quinoline from aniline and acrolein, a dehydration product of glycerol. The initial heterocyclic product of the reaction is dihydroquinoline, which must be dehydrogenated (must undergo removal of two hydrogen atoms) to give the fully aromatic product, quinoline itself.

## Ring Closure by way of Cyclic Transition States

A most important method for the synthesis of carbocyclic six-membered rings is the Diels-Alder diene reaction, named for its Nobel Prize-winning discoverers, the German chemists Otto Diels and Kurt Alder. In this reaction, illustrated below, a diene—a compound with two double bonds—reacts with a dienophile (a diene-seeking reagent), which contains a pair of carbon atoms linked by a double or triple bond. The product is a cyclohexene, a compound with a six-membered ring containing a double bond.

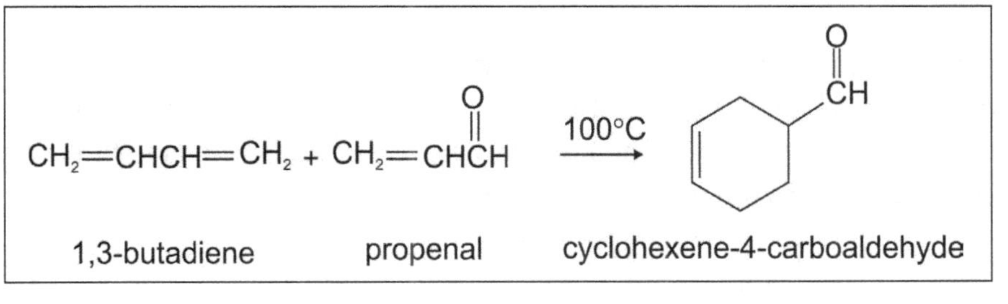


Heterocycles likewise can be synthesized by the Diels-Alder reaction, in which the dienophile contains a pair of heteroatoms such as nitrogen linked by multiple bonds.

Of even greater use, however, is a related method called the Huisgen dipolar cycloaddition reaction. This reaction is an important means of preparing many types of five-membered rings, especially those containing several heteroatoms. Pyrazoles, isoxazoles and many less-common heterocycles can be synthesized by this method.

## Conversion of one Heterocyclic Ring into Another

Although there are many reactions of theoretical importance in which one heterocyclic ring is converted into another, few are of practical use. The preparation of pyridine from tetrahydrofurfuryl alcohol and ammonia and the conversion of pyrylium salts into pyridinium salts are good examples of such transformations. In addition, ring-atom rearrangement, or "shuffling," can be brought about with light in five- and six-membered heteroaromatic compounds, and ring contraction by extrusion of an atom or a group can occur under certain conditions.

## Modification of an Existing Ring

Dehydrogenation of saturated or partially saturated heterocyclic rings to thermodynamically more-stable heteroaromatic compounds by heating with sulfur or by treatment with a palladium catalyst is analogous to similar reactions involving carbocyclic compounds.

The hydrogenation of (addition of hydrogen to) heteroaromatic rings is, by contrast, usually more difficult, for the heteroatoms tend to poison the catalyst. Finally, the modification of substituents on heterocyclic rings is of highest importance in synthesis, and reactions by which substituents may be altered are among the most useful in heterocyclic chemistry.

## Major Classes of Heterocyclic Compounds

The major classes of heterocycles containing the common heteroatoms—nitrogen, oxygen, and sulfur—are reviewed in order of increasing ring size, with compounds containing other heteroatoms. Classification by ring size is convenient because heterocyclic rings of a given size have many common features. For heterocyclic (as for carbocyclic) rings, certain broad generalizations can be made. Three- and four-membered rings, because of their small size, are geometrically strained and thus readily opened; they are also readily formed. Such heterocycles are well-known reactive intermediates. Five- and six-membered rings are readily formed and are very stable; their sizes also allow the development of aromatic character. Seven-membered rings and larger are stable but less readily formed and relatively less well investigated.

## Three-membered Rings

The three-membered ring heterocycles containing single atoms of nitrogen, oxygen, and sulfur—aziridine, oxirane (or ethylene oxide), and thiirane, respectively—and their derivatives can all be prepared by nucleophilic reactions, of the type shown. Thus, aziridine is formed by heating β-aminoethyl hydrogen sulfate with a base (in this case Y is $-OSO_3H$).

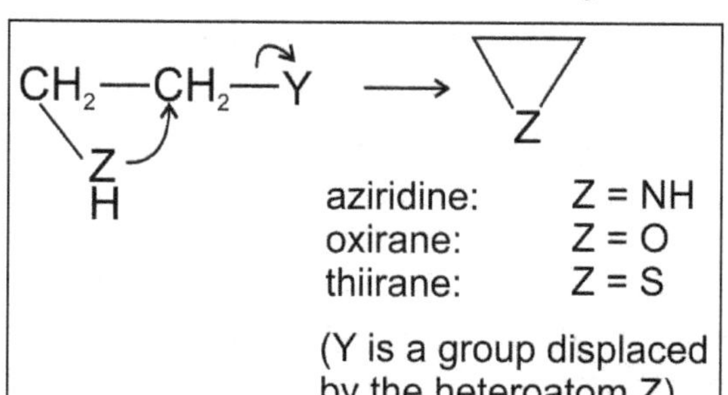

A reaction of this type is involved in the pharmacological action of nitrogen mustards, which were among the first anticancer drugs developed. Intramolecular ring closure, as in the case of the anticancer agent mechlorethamine, produces an intermediate aziridinium ion, the biologically active agent, which attacks rapidly proliferating cells such as cancer cells by inhibiting replication of their DNA (deoxyribonucleic acid). Nitrogen mustards linked to steroids also have been used as anticancer agents.

Commercially, oxirane and (to a lesser extent) aziridine are important bulk industrial chemicals. Oxirane is prepared on a large scale by the direct reaction of ethylene with oxygen.

The chemical reaction that is the most characteristic of these three-membered rings is susceptibility to attack by nucleophilic reagents to open the ring, as shown by:

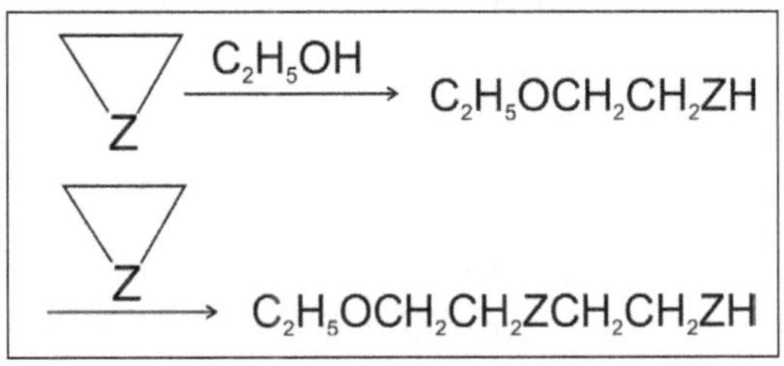


The first step shown is the reverse of the formation reaction. A s indicated, a second molecule of the three-membered ring may react with the first product. Further reaction leads to long chainlike molecules (polymers) of the type $C_2H_5O(CH_2CH_2Z)nCH_2CH_2ZH$. Polymers and copolymers of oxirane and aziridine are useful plastics.

The oxirane ring can also be opened by hydrolysis—i.e., the breaking of a bond accompanied by the addition of the elements of water (H and OH), which is similar to the first step of the ring-opening reaction illustrated above. The result of this reaction is ethylene glycol ($HOCH_2CH_2OH$), which serves as antifreeze in cooling and heating systems, in brake fluid, and as a solvent in the paints and plastics industries.

In medicine, aziridine derivatives are used for the treatment of cancer and in research as affinity probes (labeled chemicals that react selectively with biological molecules of interest) for detection and assay studies. The mitomycin family of antitumour antibiotics is the most well-known class of natural products containing the aziridine ring.

Naturally occurring compounds containing one or more oxirane rings are abundant. These molecules include insect juvenile hormones, pheromones, and products of marine organisms. Often they are biologically active and have promising therapeutic applications. For example, the fungal product (−)-ovalicin, which contains two oxirane rings, has potential for inhibiting development of solid tumours by cutting off their blood supply. An important synthetic oxirane having one ring is fosfomycin, used as an antibacterial drug, particularly in treating urinary tract infections.

Molecules containing thiirane rings are more bactericidal than those containing oxirane rings, and some thiirane derivatives have found application as tuberculostats (drugs that inhibit the growth of tuberculosis-causing bacteria), whereas thiirane 1 oxides have been reported to be insecticides, molluscicides, or herbicides.

Although aziridine, oxirane, and thiirane are saturated heterocycles, they are much more reactive than the corresponding open-chain amines, ethers, or sulfides because of the strain inherent in the three-membered ring. This behaviour reflects the comparable enhanced reactivity of the related three-membered carbocyclic compound cyclopropane.

Since 1950, five different classes of three-membered ring compounds with two heteroatoms have been discovered. They are derivatives of the parent ring systems diaziridine (containing N–N), oxaziridine (O–N), thiaziridine (S–N), dioxirane (O–O), and dithiirane (S–S). The oxathiirane system has no reported stable representatives in its class.

Except for diaziridine systems, which are used in the manufacture of plastic films and foamed plastics, as propellants and fuel stabilizers, and in the synthesis of anticancer drugs and psychopharmaceuticals, these compounds currently have limited practical importance, but they offer interesting possibilities for future applications.

## Four-membered Rings

Azetidine, oxetane, and thietane—four-membered rings containing, respectively, one nitrogen, oxygen, or sulfur atom—are prepared by nucleophilic displacement reactions similar to those used to prepare the corresponding three-membered rings.

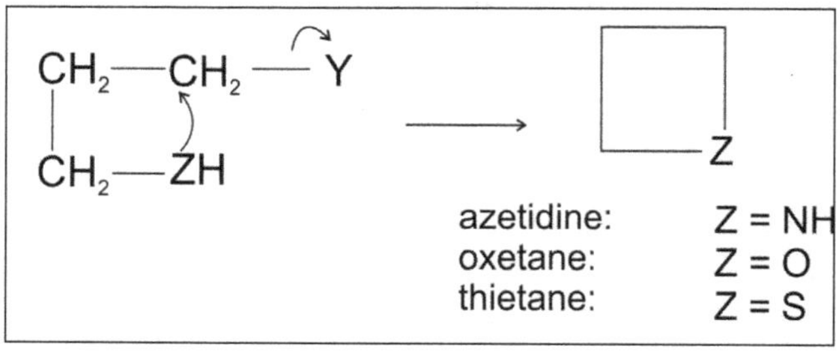


In the reaction above, Y is usually Cl, Br, or $SO_3H$. With four-membered rings, however, the reactions proceed less readily than do the analogous reactions for three-membered rings. The ring-opening reactions of four-membered heterocycles resemble qualitatively those of the corresponding three-membered rings, but they occur rather less readily.

The most important heterocycles with four-membered rings are two related series of antibiotics, the penicillins and the cephalosporins. Both series contain the azetidinone ring (the suffix -one indicating an oxygen atom linked with a double bond to a ring carbon atom). Another common name for the azetidinone ring is the β-lactam ring, which lends its name to the β-lactam antibiotics, the class to which the penicillins and cephalosporins belong. The chemistry of azetidinones was explored thoroughly during the intensive research into penicillin structure and synthesis that took place during World War II. A practical synthesis of penicillin was not achieved, however, until 1959.

Numerous oxetanes, the synthetic analogs of the antiviral natural product oxetanocin, are under investigation as antifungal, anti-inflammatory, anticancer, and antiviral agents. Oxetanones, whose ring structure is analogous to that of the azetidinones except that the heteroatom is oxygen, are widely applied in polymer manufacturing and in agriculture as herbicides, fungicides, and bactericides. The parent thietane was found in shale oil, whereas its odoriferous derivatives function as scent markers for minks, ferrets, and European polecats. Thietanes are used in the production of polymers, as bactericides and fungicides in paint, and as iron corrosion inhibitors.

## Five-membered Rings with One Heteroatom

The parent aromatic compounds of this family—pyrrole, furan, and thiophene—have the structures shown.

Ergot, a fungal disease of cereal grasses, is caused by the fungus Claviceps purpurea, which produces active chemicals containing heterocyclic compounds known as indole alkaloids.

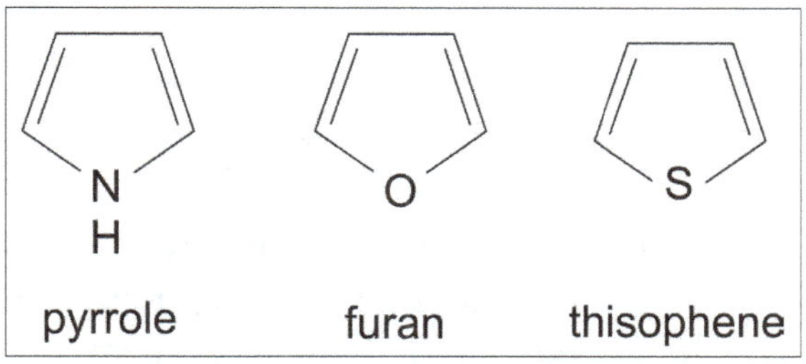


The saturated derivatives are called pyrrolidine, tetrahydrofuran, and thiophane, respectively. The bicyclic compounds made of a pyrrole, furan, or thiophene ring fused to a benzene ring are called indole (or isoindole), benzofuran, and benzothiophene, respectively.

The nitrogen heterocycle pyrrole occurs in bone oil, in which it is formed by the decomposition of proteins upon strong heating. Pyrrole rings are found in the amino acids proline and hydroxyproline, which are components of many proteins and which are present in particularly high concentrations in collagen, the structural protein of bones, tendons, ligaments, and skin.

Pyrrole derivatives are widespread in the living world. Pyrrole compounds are found among the alkaloids, a large class of alkaline organic nitrogen compounds produced primarily by plants. Nicotine is the best-known pyrrole-containing alkaloid. The heme group of the oxygen-carrying protein hemoglobin and of related compounds such as myoglobin; the chlorophylls, which are the light-gathering pigments of green plants and other photosynthetic organisms; and vitamin $B_{12}$ are all formed from four pyrrole units joined in a larger ring system known as a porphyrin, such as that of chlorophyll b, shown.

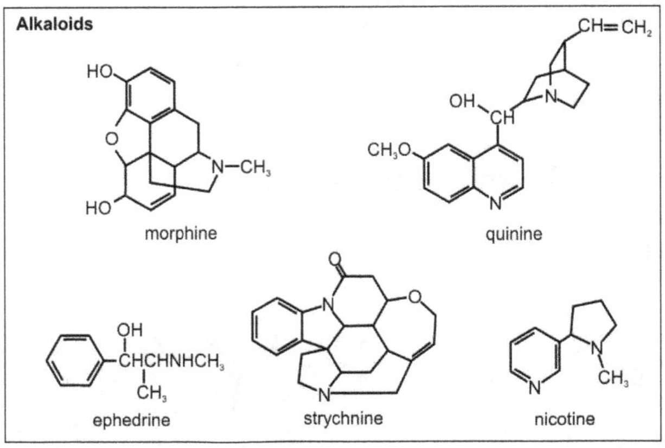


Alkaloid Chemical structures of the well-known alkaloids morphine, quinine, nicotine, strychnine, and ephedrine.

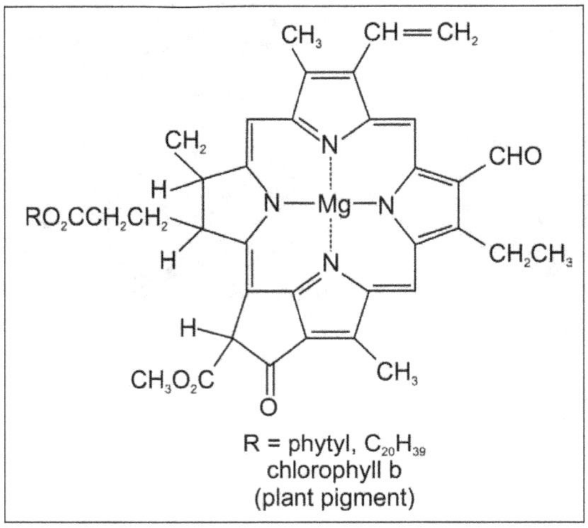


The bile pigments are formed by decomposition of the porphyrin ring and contain a chain of four pyrrole rings. Bilirubin, for example, the brownish yellow pigment that gives feces its characteristic colour, is the end product of the breakdown of heme from destroyed red blood cells.

The phthalocyanines are a group of synthetic pigments that contain four isoindole units linked together in a large ring. A typical member of the family is phthalocyanine blue (Monastral Fast Blue).

Numerous compounds produced in plants or animals contain one or more indole units in their molecular structure. The important vat dye indigo, which contains two indole units, has been used for thousands of years and was formerly obtained from plants, but it is now synthesized on a large scale.

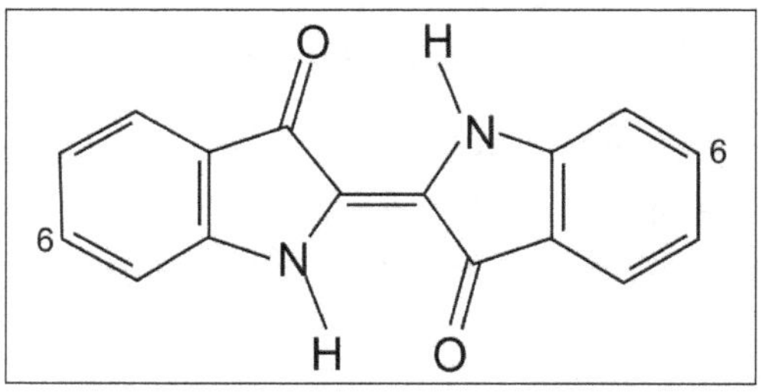

The closely related Tyrian purple, a dye obtained from species of snail and used in classical times, is 6,6′-dibromoindigo (with bromine atoms bonded to the numbered carbons in the structure).

Tryptophan, an indole-containing essential amino acid found in most proteins, is used by the body to make several important substances, including the neurotransmitter serotonin and the B-complex vitamin niacin. Skatole, a degradation product of tryptophan that retains the indole unit, contributes much of the strong odour of mammalian feces. Indole-3-acetic acid (heteroauxin or β-indolylacetic acid) is a plant-growth regulator and the most important member of the auxin family of plant hormones. The structures of these compounds are:

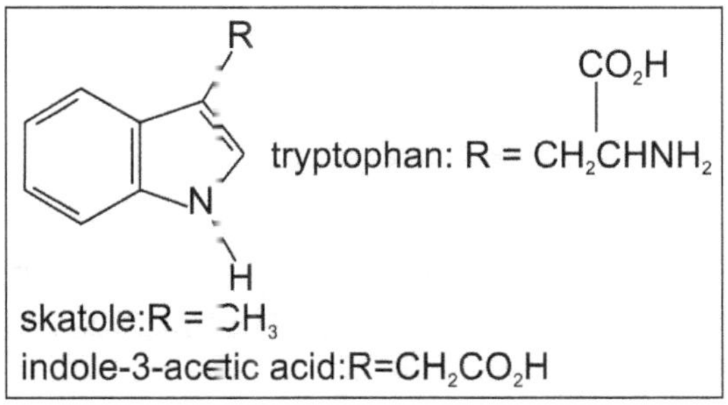


Probably the best-known indole-containing compounds are the indole alkaloids, which have been isolated from plants representing more than 30 families. The mushroom hallucinogens psilocin and psilocybin, the ergot fungus alkaloids, the drugs reserpine and yohimbine, and the poison strychnine all belong to this group.

The simplest member of the furan family of oxygen heterocycles, furan itself, is converted industrially by hydrogenation to tetrahydrofuran. Tetrahydrofuran is used as a solvent and for the production of adipic acid and hexamethylenediamine, the raw materials for nylon-6,6, the most common form of nylon. Other furan derivatives of industrial importance are maleic anhydride and phthalic anhydride, which are constituents of resins and plastics. These compounds are prepared in bulk by the oxidation of benzene and naphthalene, respectively, as shown ($V_2O_5$ is a vanadium catalyst).

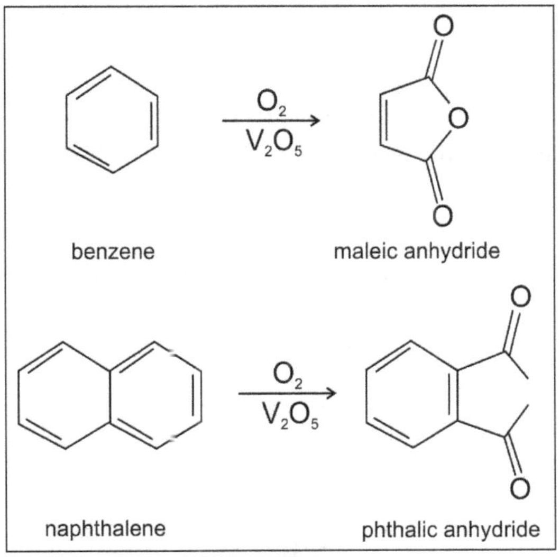


All carbohydrates, the biochemical family that includes the sugars and starches, are composed of one or more simple sugar (monosaccharide) units. These sugars are polyhydroxy aldehydes or polyhydroxy ketones that in aqueous solution exist as equilibrium mixtures of their open-chain and cyclized forms. Frequently the cyclized form of the sugar is a five-membered tetrahydrofuran ring called a furanose, as shown below for fructose, or fruit sugar, as a cyclized isomer (called a fructofuranose).

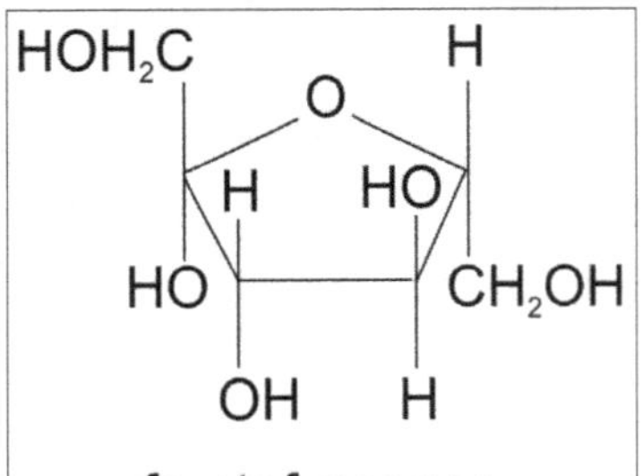

fructofuranose

Other important examples of tetrahydrofuran ring systems are the sugars ribose and deoxyribose, which are present in furanose form in, respectively, RNA and DNA, the heredity-controlling components of all living organisms.

Dehydration of certain carbohydrates yields furan derivatives. Of great commercial importance is the conversion of a carbohydrate in corncobs, oat husks, and other agricultural waste into furan-2-aldehyde, or furfural, which is used extensively as a solvent, in the manufacture of plastics, and in the preparation of other furan derivatives. Many other furan derivatives occur naturally, including vitamin C. The structures of furfural and vitamin C are:

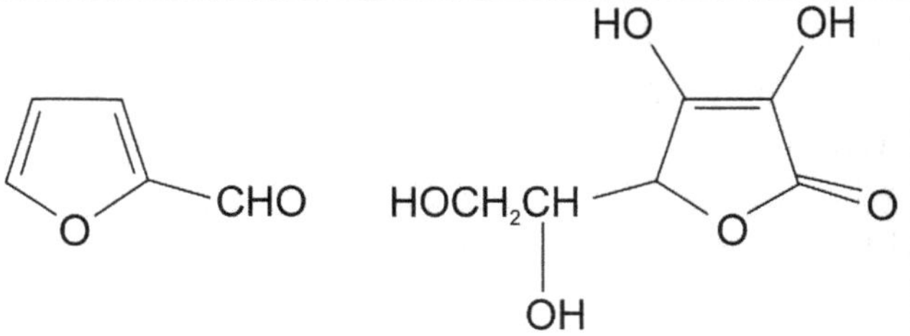

furfural

ascorbic acid (vitamin C)

The sulfur heterocycle thiophene and related compounds are found in coal tar and crude petroleum. The most important biologically occurring thiophene derivative is the B-complex vitamin biotin.

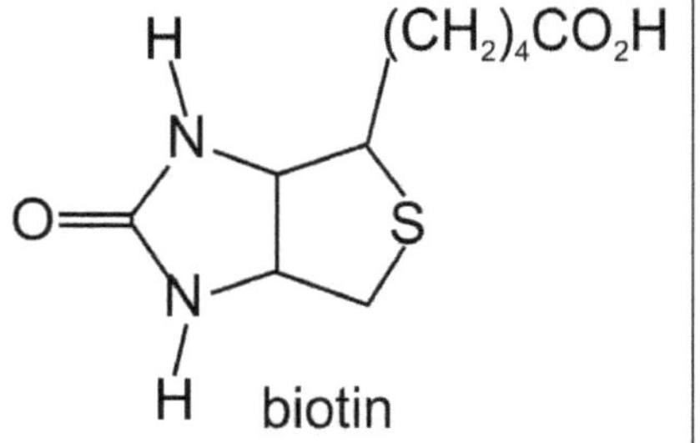

biotin (vitamin H)

## Six-membered Rings with One Heteroatom

The nomenclature used for the various monocyclic nitrogen-containing six-membered ring compounds is given below. Positions on the ring are shown for pyridine, Arabic numerals being preferred to Greek letters, although both systems are used. The pyridones are aromatic compounds because of contributions to the resonance hybrid from charged resonance forms such as that shown for 4-pyridone.

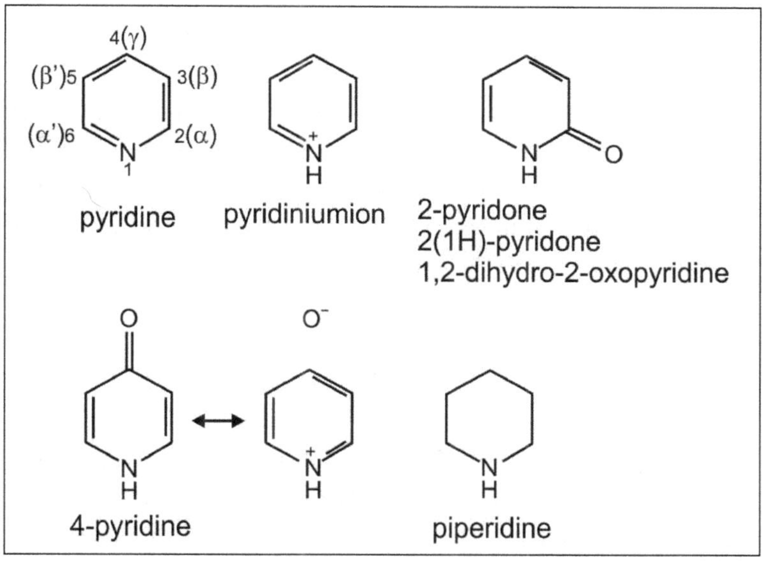


Mono-, di-, and trimethylpyridines—that is, pyridines with one, two, or three attached methyl groups, respectively—are called picolines, lutidines, and collidines, respectively, with the position of the methyl groups denoted by numbers—e.g., 2,4,6-collidine. Pyridine-2-, -3-, and -4-carboxylic acids also have widely used trivial names: picolinic, nicotinic (derived from nicotine, of which it is an oxidation product), and isonicotinic acid, respectively. Pyridine itself and the picolines, lutidines, and collidines occur in coal tar and bone oil. Pyridine derivatives are also of great biological importance. For example, nicotinic acid is more commonly known as the B-complex vitamin niacin; a nutritionally equivalent form of niacin is nicotinamide, or niacinamide. Pyridoxine is another member of the B complex, vitamin $B_6$. The structures of pyridoxine and nicotinamide are:

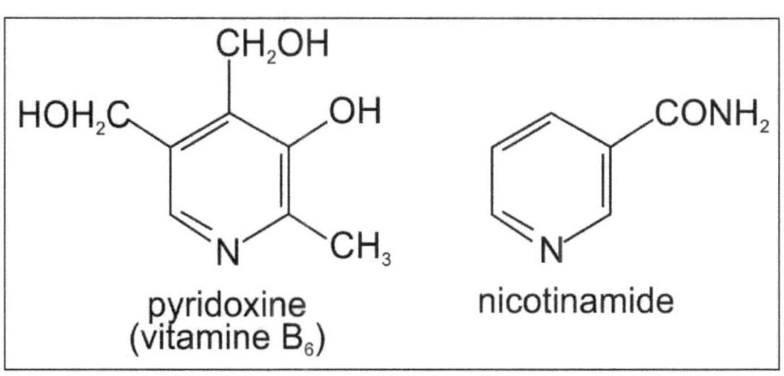


Two coenzymes involved in many important metabolic reactions in living cells, nicotinamide adenine dinucleotide (NAD, also called coenzyme I) and nicotine adenine dinucleotide phosphate (NADP, coenzyme II), are derived from nicotinamide, and the coenzyme pyridoxal phosphate (codecarboxylase) is a physiologically active form of pyridoxine. Many alkaloids contain a pyridine or piperidine ring structure, among them nicotine and piperine (one of the sharp-tasting constituents of white and black pepper, from the plant species *Piper nigrum*), with the structures shown.

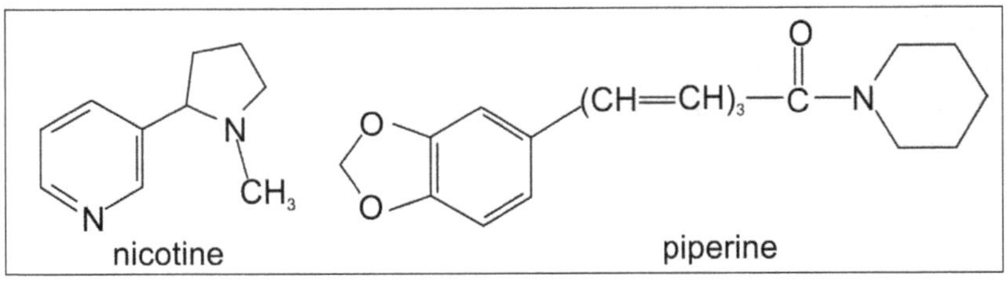


Pyridine, which once was extracted commercially from coal tar but now is prepared catalytically from tetrahydrofurfuryl alcohol and ammonia, is an important solvent and intermediate used to make other compounds. Vinylpyridines are important monomer building blocks for plastics, and fully saturated pyridine, piperidine, is used in rubber processing and as a chemical raw material.

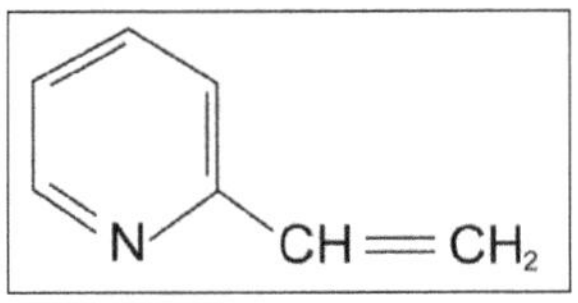


Pharmaceutically important pyridines include the tuberculostat isoniazid (isonicotinic acid hydrazide), the anti-AIDS-virus drug nevirapine, the vasodilator nicorandil, used for treating angina, the urinary-tract analgesic phenazopyridine, and the anti-inflammatory sulfa drug. 1-(1-phenylcyclohexyl) piperidine (PCP, phencyclidine) was originally used as an anesthetic, but its powerful hallucinogenic properties have led to abuse. Diquat, paraquat, clopyralid, and diflufenican are well-known pyridine derivatives used as herbicides.

Shown in the structural formulas below are two isomeric benzopyridines (upper pair) and two isomeric dibenzopyridines (lower pair), with their common names and accepted numberings. All four compounds and some of their alkyl derivatives have been obtained from coal tar. Each of them is also the parent substance of a class of alkaloids. Of these, the quinolines (e.g., quinine and other derivatives, still obtained from the Cinchona tree) and the isoquinoline (e.g., morphine) groups are particularly well-known.

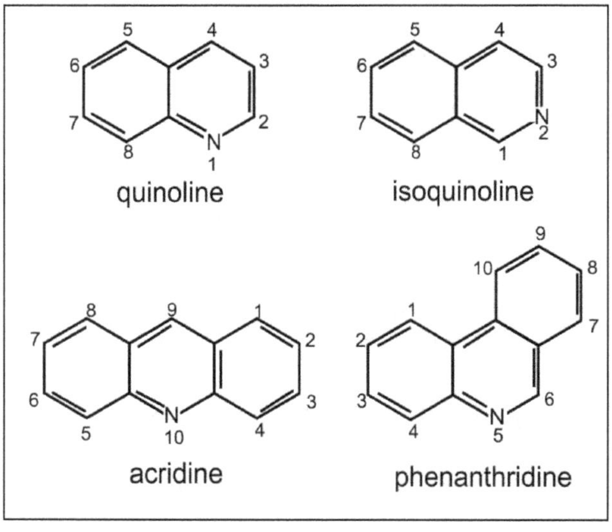


Quinoline itself can be used to manufacture nicotinic acid and other compounds such as drugs and dyes. The production of synthetic quinoline far exceeds that from coal tar.

Morphine, codeine, and thebaine—all containing partially reduced isoquinoline rings—are alkaloids of the opium poppy and have been used for many centuries as hypnotics and analgesics. The semisynthetic derivative of morphine, heroin, is an even more powerful hypnotic and a highly addictive drug.

Important synthetic derivatives of the benzo- and dibenzopyridines include cyanine dyes, used as sensitizers in silver halide photographic emulsions, and quinophthalone dyes, applied in plastics, polymer textiles, paper, cosmetics, transfer-printing processes, electronic photography, and laser dyes. Other derivatives are useful bacteria-staining agents for microscopy, antiseptics such as the coal-tar dye acriflavine, antimalarial agents such as mepacrine (quinacrine) and chloroquine, antibacterial agents such as ciprofloxacin, trypanocides (drugs that destroy trypanosomes, which are parasitic protozoans responsible for Chagas disease, sleeping sickness, and other serious infectious illnesses) such as ethidium bromide, and the reagent oxine (8-hydroxyquinoline or 8-quinolinol), used in analytical chemistry.

Positively charged ions (cations) of pyrylium and thiopyrylium are the parent six-membered, aromatic, monocyclic oxygen and sulfur compounds of their respective groups.

pyrylium oation thiopyrylium oation

An uncharged aromatic (completely conjugated) six-membered ring containing an oxygen or sulfur atom is possible only if the ring contains a carbonyl group (i.e., a ring carbon atom linked by a double bond to an oxygen atom not in the ring), as in the pyrones.

The pyrans contain extra hydrogen atoms, the position of which is indicated in structural diagrams by a number followed by an *H*. Certain sugars—a typical example is the monosaccharide glucose—are called pyranoses because they contain six-membered tetrahydropyran rings, with the structure:

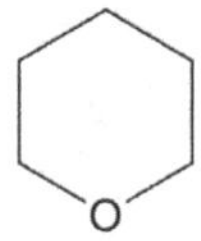

tetrahydropyran

Pyrone derivatives are present in natural products. Kojic acid, for example, is an antibiotic derived by action of certain molds on starches or sugars. The steroid bufotalin and its poisonous ester bufotoxin are obtained from the skin glands of toads.

The benzopyrylium cation is the parent of a large number of natural products. Chroman, or 3,4-dihydro-2*H*-1-benzopyran, is itself not found in nature, but the chroman unit is present in many natural products. Vitamin E (α-tocopherol), a substituted chroman, is found in plant oils and the leaves of green vegetables, whereas coumarin, or 2*H*-1-benzopyran-2-one, used in perfumes and flavourings, and its derivative dicoumarin (dicumarol, or discoumarol), a blood anticoagulant, are products of living organisms.

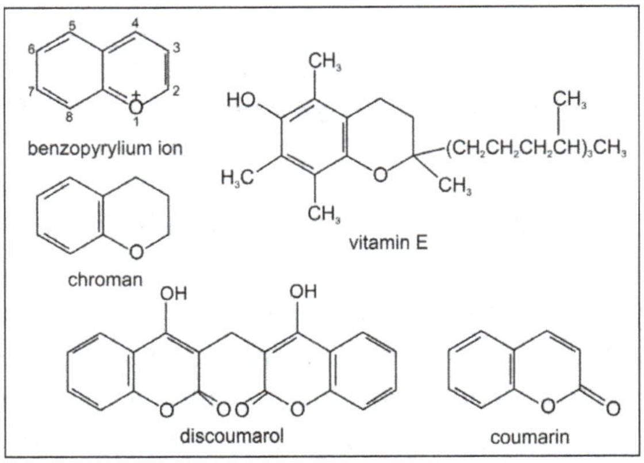


The flavylium cation is the parent of the anthocyanidines, substances that in chemical combination with sugars form the anthocyanin pigments, the common red and blue colouring matters of flowers and fruits. The colour range from yellow to reddish orange is provided by anthoxanthins, which constitute a subgroup of flavonoids. The latter are derivatives of flavone (2-phenyl-4-pyrone) in which one or more hydrogen atoms are replaced by hydroxy or methoxy groups. The flavylium ion has the structure:

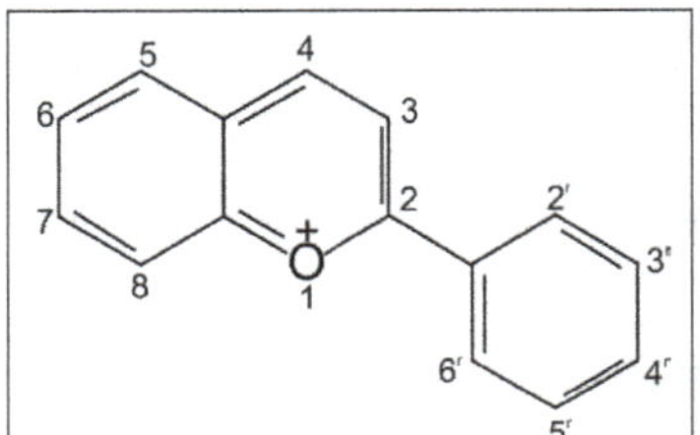

flavylium ion

## Five- and Six-membered Rings with Two or more Heteroatoms

The names and numbering systems for the five-membered heteroaromatic rings with two heteroatoms are:

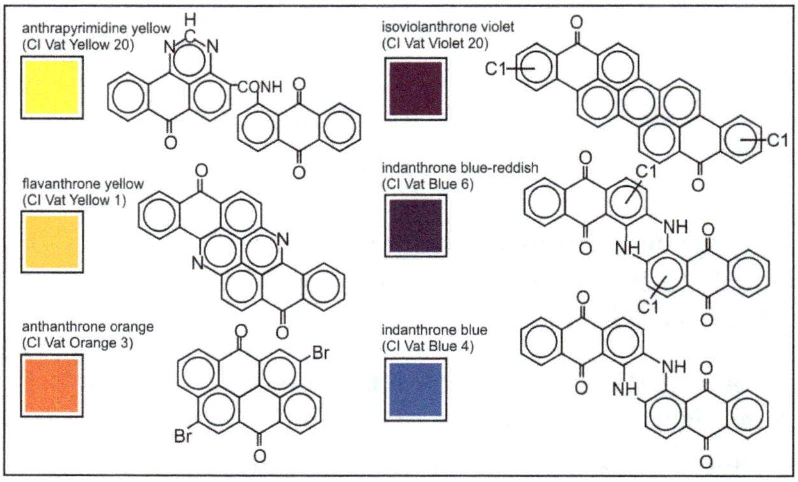


Anthrapyrimidine yellow, flavanthrone yellow, indanthrone blue-reddish, and indanthrone blue are examples of heterocyclic anthraquinone dyes.

Few pyrazoles occur naturally; the compounds of this class are usually prepared by the reaction of hydrazines with 1,3-diketones. Many synthetic pyrazole compounds are of importance as dyes and medicinals. Among them are the fever-reducing analgesic aminopyrine, the anti-inflammatory drug phenylbutazone, used in treating arthritis, the yellow food colour and fibre dye tartrazine, and a series of dyes used as sensitizing agents in colour photography.

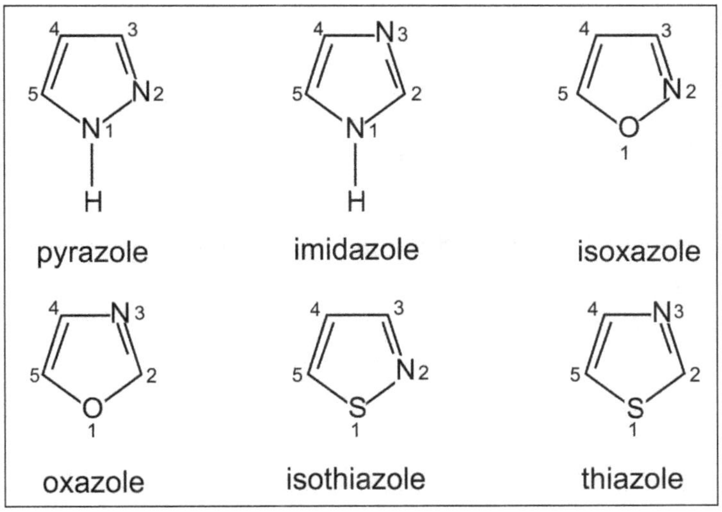


Imidazoles are most important biologically; histidine, for example, is an essential amino acid of particular importance in enzyme reactions. A breakdown product of histidine, called histamine, has a variety of functions in different organisms; in the human body it plays a crucial role in the immune response, including allergic reactions—hence the importance of antihistamine drugs. Histidine and histamine have the structures:

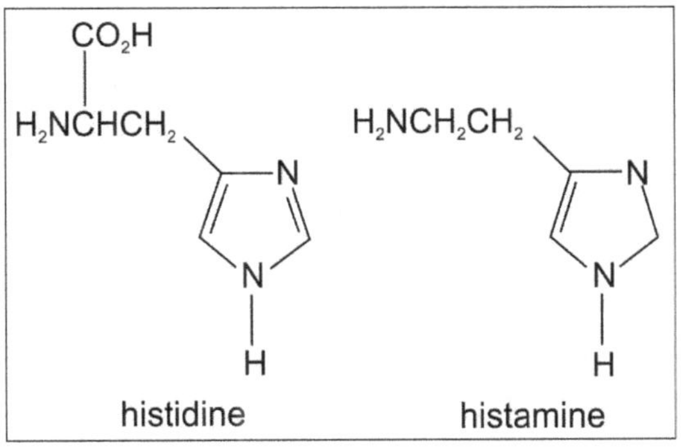


Among other naturally occurring compounds with an imidazole nucleus are hydantoin, which is found in beet sap, and allantoin, which is related to the metabolic product uric acid. Hydantoin derivatives, in particular phenytoin, are important antiepileptic drugs. The imidazole ring is also present in the B vitamin biotin.

The antibiotic cycloserine, produced by a bacterium, is one of the few naturally occurring isoxazoles. A synthetic isoxazole, hymexazol, has found practical use as soil and seed fungicide.

Thiazoles are of great biological importance. This ring system occurs in thiamin (thiamine, vitamin $B_1$), the bacitracin and penicillin antibiotics (from a bacterium and a mold, respectively), and in numerous synthetic drugs, dyes, and industrial chemicals. Synthetic drugs belonging to the thiazole family include the antimicrobial agents sulfathiazole and acinitrazole, the antidepressant pramipexole, and the antiasthmatic drug cinalukast. Sulfathiazole has the structure:

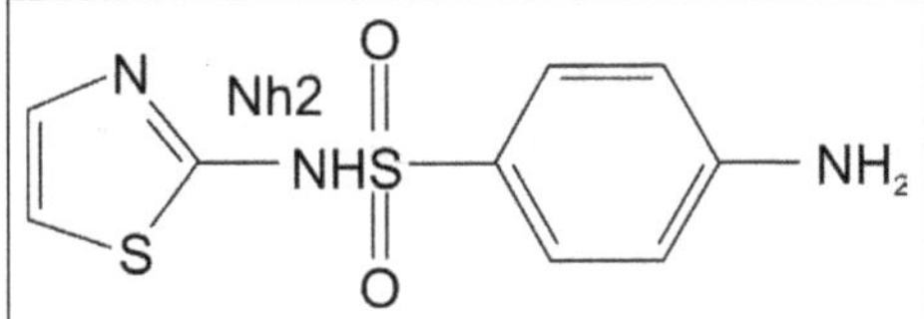

sulfathiazole

Other thiazole compounds include rhodanine, the dye rhodanine red derived from it, and the yellow dye primuline.

Most bicyclic systems derived from these five-membered rings are named systematically—that is, by use of the prefix benz- or benzo- to indicate the presence of the benzene ring. Benzimidazole, for example, is the name for the compound:

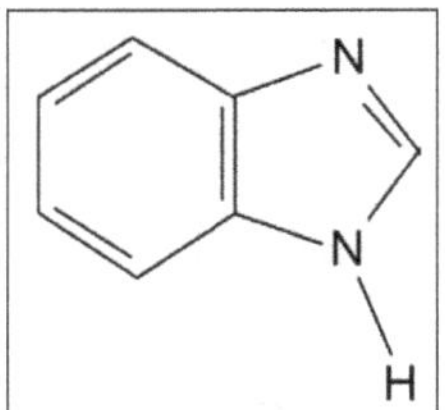

benzimidazole

A benzimidazole unit occurs in vitamin $B_{12}$. Benzothiazole derivatives are used for accelerating the vulcanization of rubber (2-mercaptobenzothiazole), as herbicides (benazolin, mefenacet), and as fungicides and antihelminthic drugs (thiabendazole).

The three monocyclic diazines—six-membered ring compounds with two nitrogen heteroatoms—are named and numbered as shown.

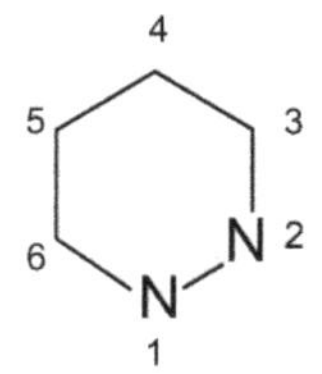
pyridazine

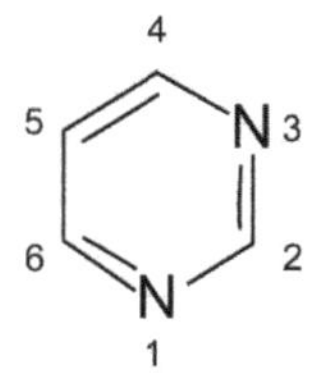
pyrimidine

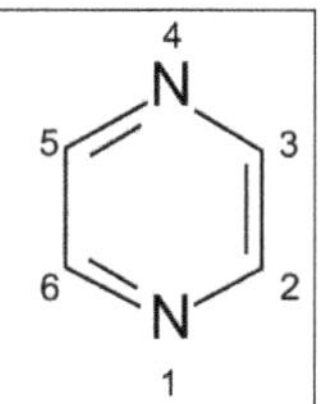
pyrazine

The pyridazine derivative maleic hydrazide is a herbicide, and some pyrazines occur naturally—the antibiotic aspergillic acid, for example. The structures of the aforementioned compounds are:

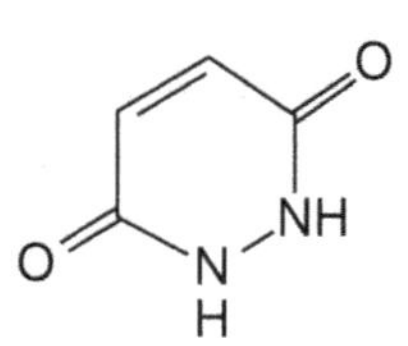

maleic hydrazide

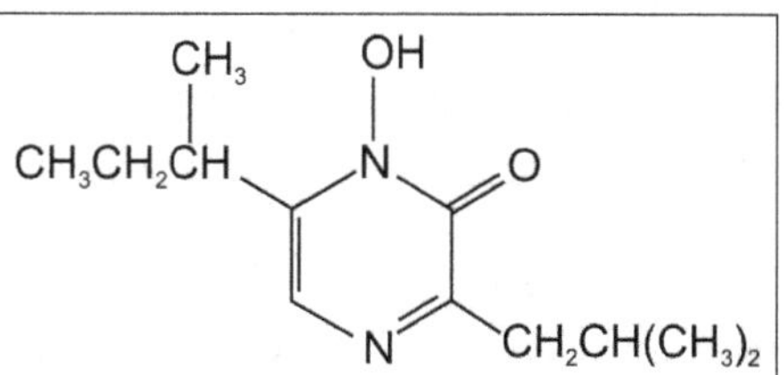

aspergillic acid

The pyrazine ring is a component of many polycyclic compounds of biological or industrial significance. Important members of the pyrazine family include pteridines, alloxazines, and phenazines.

Biologically and pharmacologically, however, the most important diazines are the pyrimidines. Uracil, thymine, and cytosine, for example, with the structures shown, are three of the five nucleotide bases that constitute the genetic code in DNA and RNA.

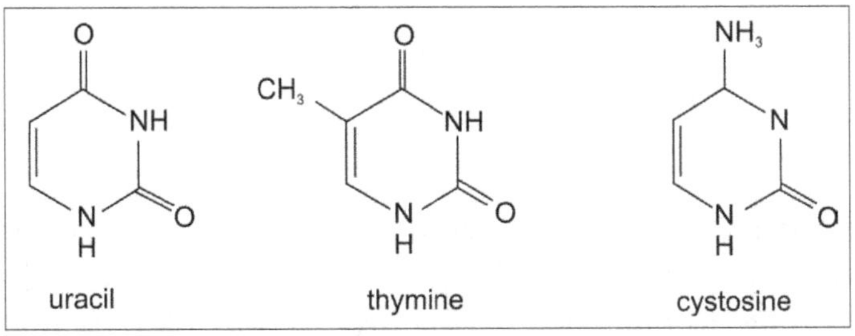


The vitamin thiamin contains a pyrimidine ring (in addition to the five-membered thiazole ring, and synthetic barbiturates such as amobarbital (amylobarbitone) are widely used drugs.

Various oxazine and thiazine derivatives are known, but monocyclic thiazines are as yet of little importance. The parent tetrahydro-1,4-oxazine, commonly called morpholine, is produced on a large scale for use as a solvent, corrosion inhibitor, and fungicide. The morpholine ring is also present in the sedative-hypnotic drug trimetozine and in some fungicides such as tridemorph and fenpropimorph. The structural formula for morpholine is:

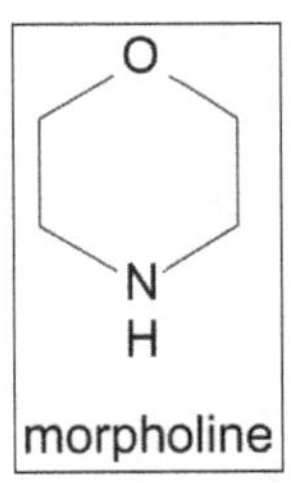


The benzodiazines are polycyclic compounds containing one or more benzene rings fused to a diazine ring. Many have common names—e.g., cinnoline, quinazoline, and phenazine.

Five other polycyclic systems in this general family are of significance, their names, structures, and numbering systems being:

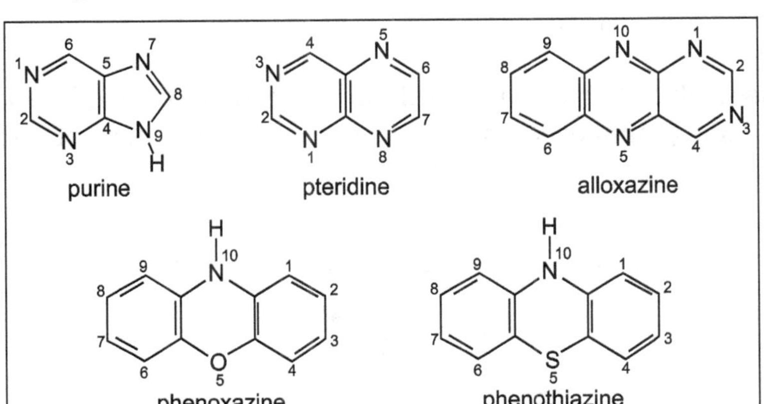

Quinoxaline alkaloids exist, and there are some phenazine natural products. Phenazine dyes are used for fabrics and for inks and printer toners (nigrosines). The first synthetic phenazine dye, mauve (aniline purple), is historically important; its structure is:

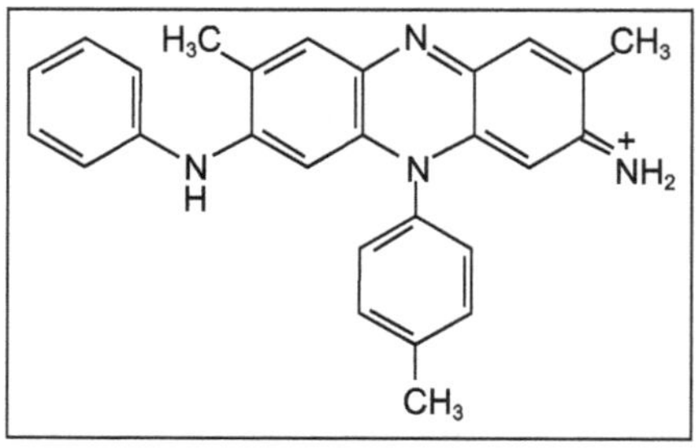

The phenoxazine system is a chromophoric (colour-imparting) part of the molecular structures of the naturally occurring actinomycin antibiotics, which are yellow-red in colour. Many polycyclic compounds containing a phenoxazine ring are used as biological stains, fabric dyes, and light-emitting materials in dye lasers (e.g., cresyl violet and nile blue).

Phenothiazine has been used as a worming agent for livestock and as an insecticide. Drugs of the phenothiazine type include the antipsychotic agents chlorpromazine and thioridazine, the long-acting antihistamine promethazine, and ethopropazine, used in treatment of parkinsonism. A large group of dyes have the phenothiazine structure, including methylene blue, a substance widely used as a biological stain and an oxidation-reduction indicator. The structure of methylene blue is:

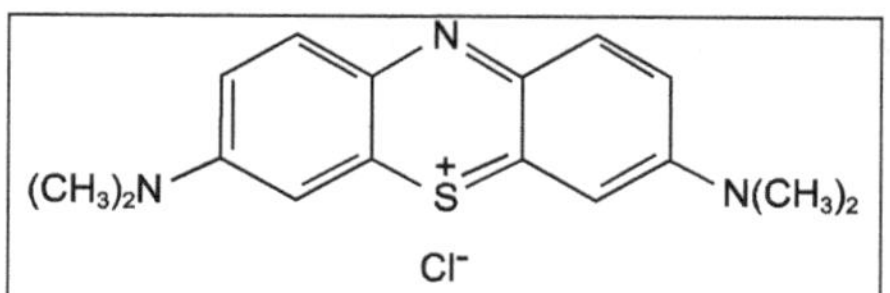

methylene blue

Biologically, the purines and pteridines are the most important polycyclic diazines. Purine itself is not common, but the purine structure is present in many natural substances. Two purine nucleotide bases, adenine and guanine, occur together with the pyrimidine bases in DNA and RNA mentioned above.

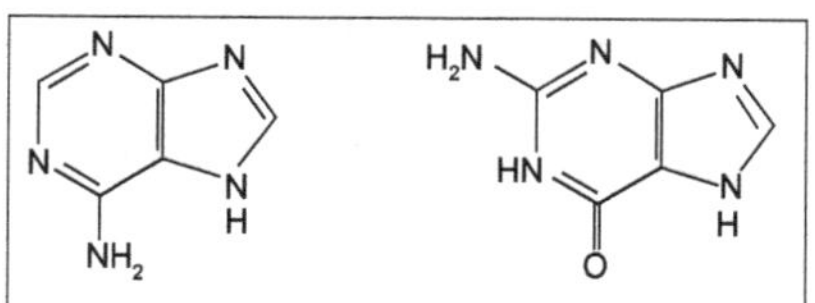

adenine guanine

Other natural purines include the alkaloids xanthine and caffeine (found in tea, coffee, and cocoa), the alkaloid theobromine (found in cocoa), and uric acid. The structures of caffeine, theobromine, and uric acid are:

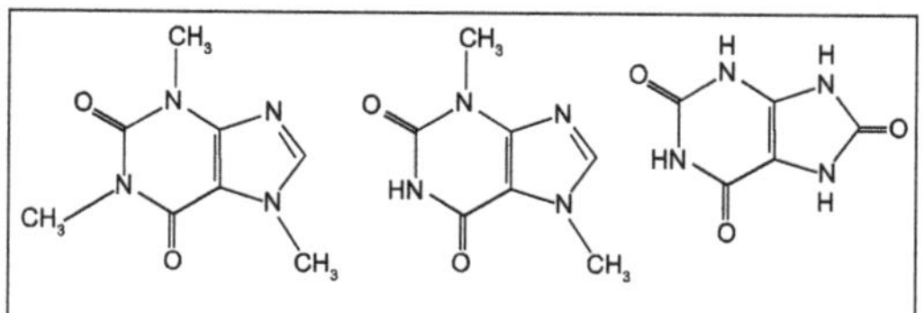

caffeine theobromine uric acid

Adenosine monophosphate, diphosphate, and triphosphate (AMP, ADP, and ATP, respectively) are important participants in energy processes in the living cell. Each of the compounds is composed of the nucleotide base adenine linked to the sugar ribose, which in turn is linked to a linear "tail" of one, two, or three phosphate groups, respectively, as shown.

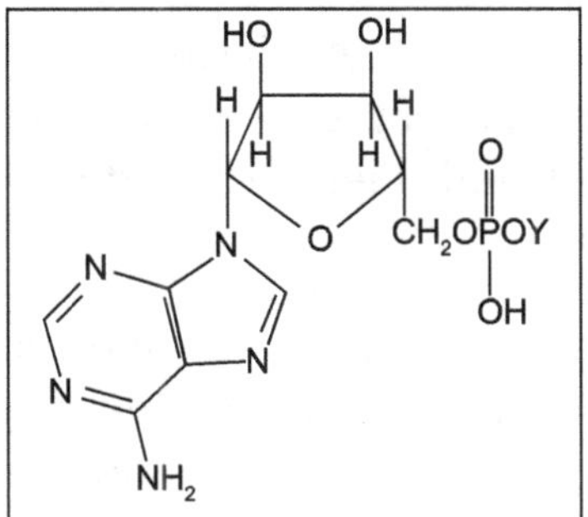


adenosine monophosphate: Y = H

adenosine diphosphate: Y = $PO_3H_2$

adenosine triphosphate: Y = $P_2O_6H_3$

The biological significance of pteridine compounds has become apparent since the first known members of the group were discovered as pigments of butterfly wings. One example is the yellow pigment 2-amino-4,6-pteridinedione (xanthopterin).

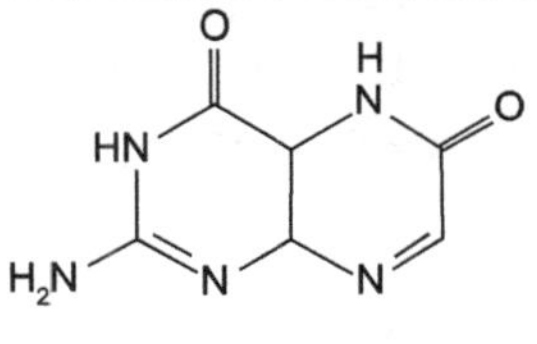


2-amino-4,6-pteridinedione

Folic acid, also a pteridine, is a B-complex vitamin and an important growth factor.

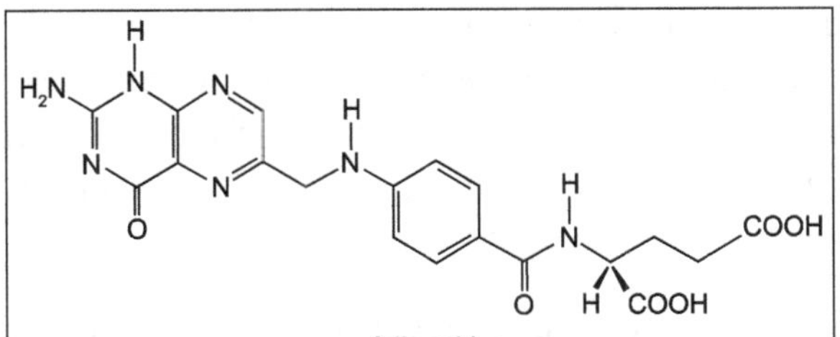


folic acid

Riboflavin, or vitamin $B_2$, is a derivative of alloxazine.

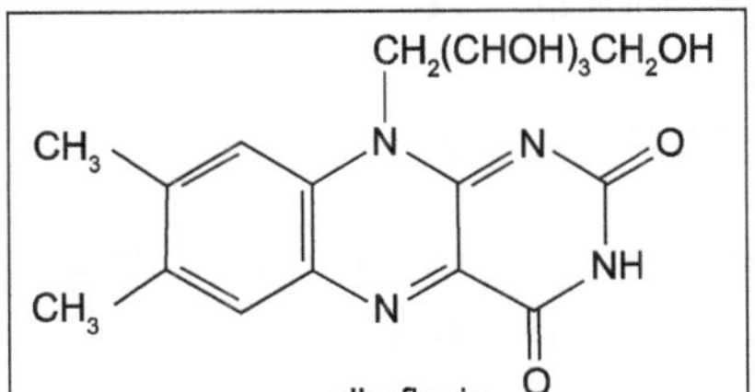


riboflavin

## Rings with Seven or more Members

As the size of the ring increases, the range of compounds that can be obtained by varying the number, type, and location of the heteroatoms increases enormously. Nevertheless, the chemistry of heterocyclic compounds with rings seven-membered or larger is much less developed than that of five- and six-membered ring heterocycles, although these compounds are usually stable and some of them have found practical application. Of the seven-membered ring compounds, one-heteroatom heterocycles—azepines, oxepines, and thiepines—and their derivatives are the most comprehensively studied.

The increase in ring size constrains these compounds to be nonplanar in order to lessen the ring strain. Nonplanarity, however, affects aromaticity, so these heterocycles react as cyclic polyenes (compounds with noninteracting, alternating single and double bonds). Azepine and oxepine rings are important constituents of numerous naturally occurring alkaloids and metabolic products of marine organisms. The azepine derivative caprolactam is produced commercially in bulk for use as an intermediate in the manufacture of nylon-6 and in production of films, coatings, and synthetic leather. Seven-membered heterocycles with one or two nitrogen atoms in the ring are structural units of widely used psychopharmaceuticals such as imipramine (trade name Prazepine)—the first of the tricyclic antidepressants—and the tranquilizer diazepam (trade name Valium).

Of the larger ring heterocycles, the most important are the crown ethers, which contain one or more heterocyclic rings comprising 12 or more ring atoms and involving a number of various heteroatoms, usually nitrogen, oxygen, or sulfur. The heteroatoms are usually separated by two-carbon or three-carbon units (ethylene or propylene units, respectively). The first crown ether, dibenzo-18-crown-6, was synthesized in 1960.

The first number in a crown ether name indicates the total number of atoms involved in the macrocycle (i.e., the large ring), while the second indicates the number of heteroatoms in that ring. The remarkable feature of crown ethers, which stimulated the explosive development of their chemistry, is their ability to selectively bind the ions of metal elements (e.g., potassium and sodium) and whole organic molecules inside their cavities, the selectivity for a particular ion or molecule being directly related to the size of the macrocycle. Because of this feature, crown ethers have found wide application as ion transporters, as materials for ion-selective electrodes used in environmental testing for various metal ions, as sensitizers in photography, in medical diagnostics, and for the separation of radioactive isotopes.

Although crown ethers are not found in nature, some larger ring heterocycles that possess similar pronounced binding abilities exist as natural products. An example is the porphyrins, which are widely distributed as biological pigments—e.g., the magnesium-binding chlorophylls and the iron-binding heme groups of hemoglobin and myoglobin.

## Rings with Uncommon Heteroatoms

In addition to the nitrogen, oxygen, and sulfur atoms commonly found in heterocycles, a large number of other elements form such rings—of greater or lesser stability. Such compounds are as yet of little practical importance.

## Halogens, Selenium and Tellurium

Cyclic chloronium, bromonium, and iodonium ions—ions of the halogen elements chlorine, bromine, and iodine, respectively—have been prepared. Of these, only the iodine derivative has much stability.

Many heterocycles containing selenium (Se) atoms are known. Selenium shows much similarity in behaviour to sulfur; hence, selenophene, with the structure shown, resembles thiophene quite closely.

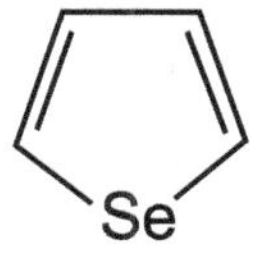


selenophene

Carbon-selenium and selenium-selenium bonds are considerably weaker than the corresponding carbon-sulfur and sulfur-sulfur bonds; consequently, the reactions of selenium compounds tend to involve the heteroatom.

Tellurium (Te) heterocycles are rarer and even less stable than selenium heterocycles. One of the first such compounds, prepared in 1971, is dibenzotellurophene.

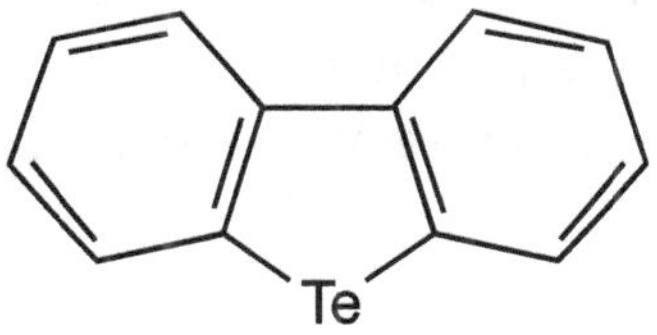


dibenzotellurophene

Five-membered selenium heterocycles, such as selenophene-6 (featuring both selenophene and pyridine rings) and benzoselenophenes (selenium analogs of benzothiophenes, play an important role as antioxidants in living organisms. Antioxidants reduce damage done to cells by free radicals, highly reactive molecules that are released during normal metabolic processes. Because free radicals can also be produced by exposure to ionizing radiation, these natural selenium compounds constitute potential radioprotective agents. The enhanced toxicity of the majority of selenium compounds compared with their sulfur analogs, however, significantly restricts the medical applications of these compounds.

In industry, selenium heterocycles are used almost exclusively in the preparation of "organic metals"—organic materials that have high electrical conductivity, much like metals.

## Phosphorus, Arsenic, Antimony and Bismuth

Phosphorus (P), arsenic (As), antimony (Sb), and bismuth (Bi), which are all members of group Va of the periodic table of elements, form a closely related group of heterocycles. There is, however, little similarity between their properties and those of the corresponding derivatives of nitrogen, a member of the same group. Although phosphorus-containing heterocycles have long been known, the heterocyclic chemistry of arsenic, antimony, and bismuth has made significant progress only in recent years because of the decreased stability of heterocycles involving a heavy element. Parent five-membered heterocycles with arsenic, antimony, or bismuth have yet to be isolated, and six-membered bismuth heterocycles are unknown. Because of this instability, those antimony and bismuth compounds that have been synthesized have found little practical application.

Although many organophosphorus compounds are used in medicine or as insecticides and herbicides, they include few phosphorus-containing heterocycles. Two phosphorus heterocycles of practical importance are the anticancer drug cyclophosphamide and the insect chemosterilant apholate; the latter compound also contains six three-membered aziridine rings. The arsenic heterocycle 10,10′-oxybisphenoxarsine is used as a fungicidal and bactericidal additive to plastics.

## Silicon, Germanium, Tin and Lead

Although silicon (Si), germanium (Ge), tin (Sn), and lead (Pb) fall in the same periodic group (IVa) as carbon, they demonstrate much lesser ability to form stable chains or double bonds. Nevertheless, chemists know of many saturated, unsaturated, and even aromatic heterocycles that incorporate one or more atoms of these elements into their ring systems. The traditional method for preparing various such compounds is based on coupling of the halogen derivatives of the type $R_2MCl_2$, in which M is silicon, germanium, tin, or lead and R is halogen or a hydrocarbon chain, with organomagnesium or organolithium reagents.

All four of these elements tend to form stronger bonds with nitrogen and, especially, with oxygen than with carbon. Alternating replacement of all the carbon in carbocyclic compounds with silicon and oxygen results in silicates, which are inorganic analogs of organic heterocycles.

## Boron

A large variety of heterocycles with five-, six-, or seven-membered rings containing boron (B) have been prepared and studied. Several saturated boron heterocycles were found to be more stable than their open-chain analogs, suggesting that the boron-containing cyclic structure itself favours stability. One of the best-known saturated heterocyclic boranes, widely used in organic chemistry as a reducing agent, is 9-borabicyclo[3.3.1]nonane (9-BBN).

A boron atom and a nitrogen atom together contain the same number of electrons as two carbon atoms. Not surprisingly, a boron-nitrogen unit can replace a carbon-carbon unit in benzenoid compounds to give stable heteroaromatics. A good example is 9-aza-10-boraphenanthrene, which incorporates a boron and a nitrogen atom linked by a double bond in one of its rings.

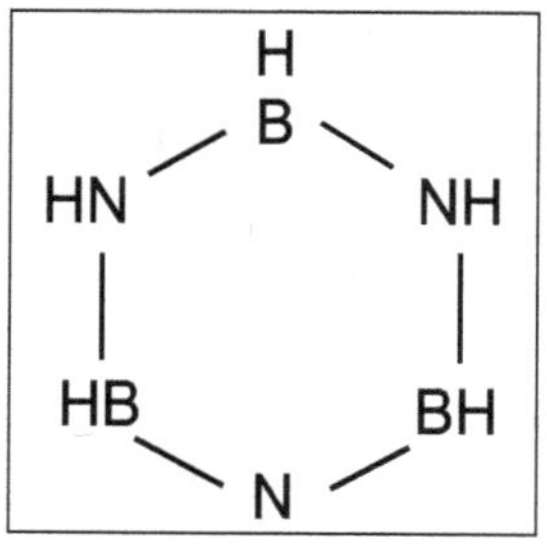


The exhaustive substitution of all two-carbon units in a cyclohexane ring or a benzene ring with alternating boron-nitrogen units produces borazine or borazole, respectively; the latter is often referred to as inorganic benzene.

# Organosulfur Compound

Organosulfur compounds are organic compounds that contain sulfur. They are often associated with foul odors, but many of the sweetest compounds known are organosulfur derivatives, e.g., saccharin. Nature abounds with organosulfur compounds—sulfur is essential for life. Of the 20 common amino acids, two (cysteine and methionine) are organosulfur compounds, and the antibiotics penicillin and sulfa drugs both contain sulfur. While sulfur-containing antibiotics save many lives, sulfur mustard is a deadly chemical warfare agent. Fossil fuels, coal, petroleum, and natural gas, which are derived from ancient organisms, necessarily contain organosulfur compounds, the removal of which is a major focus of oil refineries.

Sulfur shares the chalcogen group with oxygen, selenium, and tellurium, and it is expected that organosulfur compounds have similarities with carbon–oxygen, carbon–selenium, and carbon–tellurium compounds.

A classical chemical test for the detection of sulfur compounds is the Carius halogen method.

## Classes

Organosulfur compounds can be classified according to the sulfur-containing functional groups, which are listed (approximately) in decreasing order of their occurrence.

### Illustrative Organosulfur Compounds

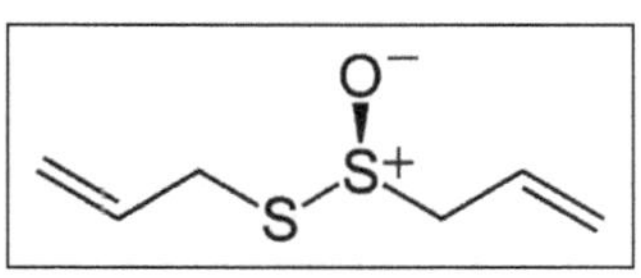

Allicin, the active flavor compound in crushed garlic

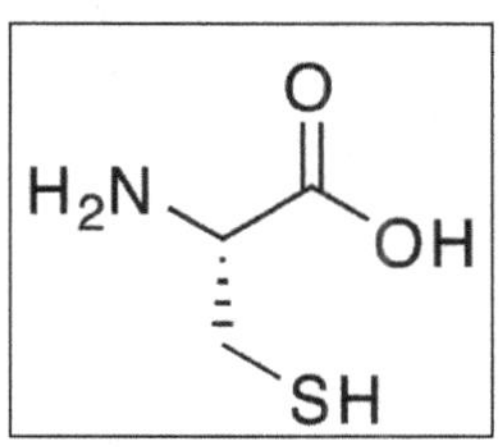

(*R*)-Cysteine, an amino acid containing a thiol group

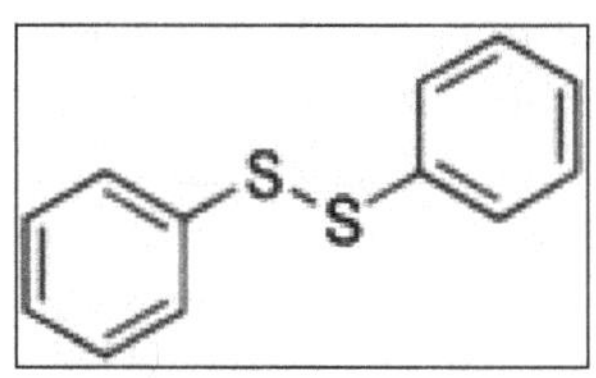

Diphenyl disulfide, a representative disulfide

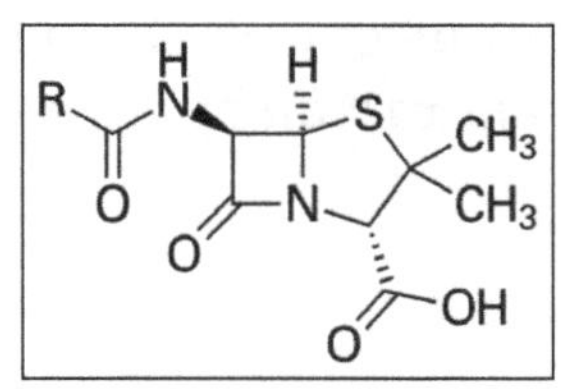

Penicillin core structure, where "R" is the variable group.

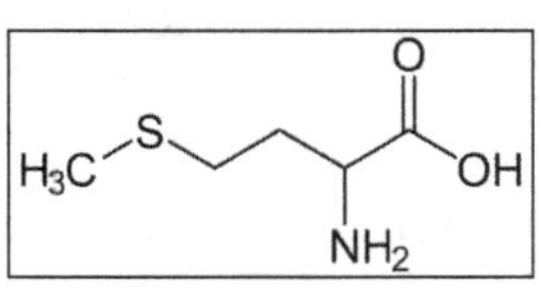

Methionine, an amino acid containing a sulfide

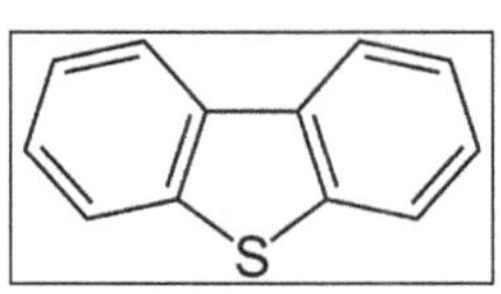

Dibenzothiophene, a component of crude oil

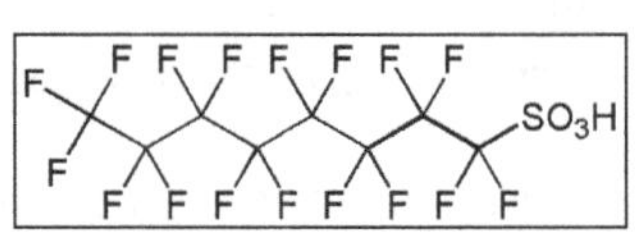

Perfluorooctanesulfonic acid, a controversial surfactant

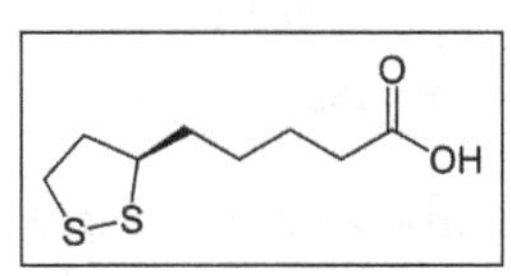

Lipoic acid, an essential cofactor of four mitochondrial enzyme complexes.

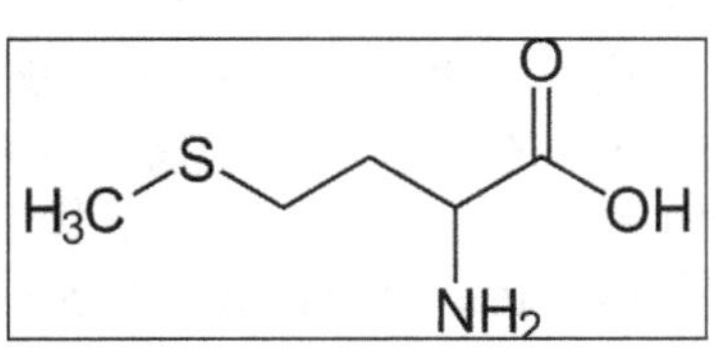

Sulfanilamide, a sulfonamide antibacterial, called a sulfa drug.

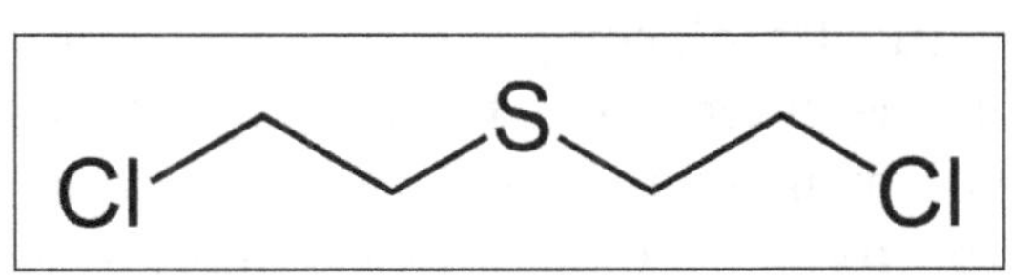

Sulfur mustard, a type of sulfide used as a chemical warfare agent.

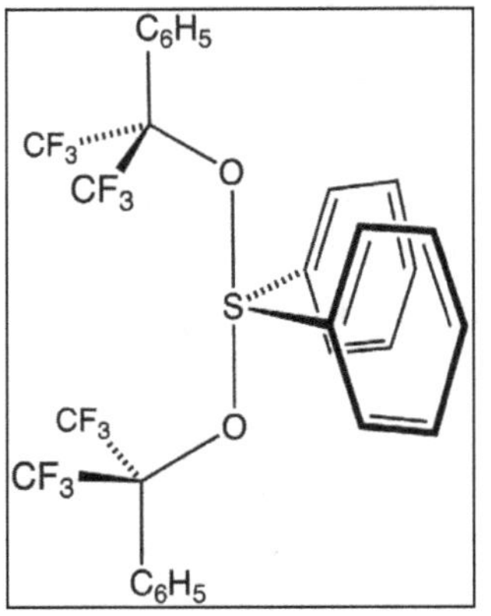

Martin's sulfurane with a see-saw structure, like that of $SF_4$.

## Sulfides

Sulfides, formerly known as thioethers, are characterized by C–S–C bonds Relative to C–C bonds, C–S bond are both longer, because S is larger than carbon, and about 10% weaker. Representative bond lengths in sulfur compounds are 183 pm for the S–C single bond in methanethiol and 173 pm in thiophene. The C–S bond dissociation energy for thiomethane is 89 kcal/mol (370 kJ/mol) compared to methane's 100 kcal/mol (420 kJ/mol) and when hydrogen is replaced by a methyl group the energy decreases to 73 kcal/mol (305 kJ/mol). The single carbon to oxygen bond is shorter than that of the C–C bond. The bond dissociation energies for dimethyl sulfide and dimethyl ether are respectively 73 and 77 kcal/mol (305 and 322 kJ/mol).

Sulfides are typically prepared by alkylation of thiols. They can also be prepared via the Pummerer rearrangement. In one named reaction called the Ferrario reaction phenyl ether is converted to *phenoxathiin* by action of elemental sulfur and aluminium chloride.

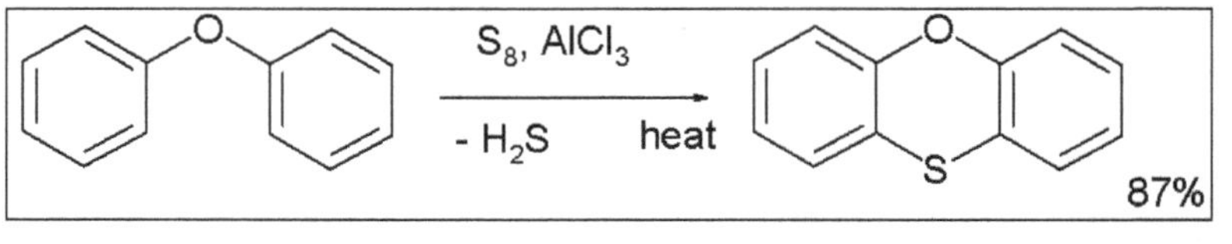


Thioacetals and thioketals feature C–S–C–S–C bond sequence. They represent a subclass of sulfides. The thioacetals are useful in "umpolung" of carbonyl groups. Thioacetals and thioketals can also be used to protect a carbonyl group in organic syntheses.

The above classes of sulfur compounds also exist in saturated and unsaturated heterocyclic structures, often in combination with other heteroatoms, as illustrated by thiiranes, thiirenes, thietanes, thietes, dithietanes, thiolanes, thianes, dithianes, thiepanes, thiepines, thiazoles, isothiazoles, and thiophenes, among others. The latter three compounds represent a special class of sulfur-containing heterocycles that are aromatic. The resonance stabilization of thiophene is 29 kcal/mol (121 kJ/mol) compared to 20 kcal/mol (84 kJ/mol) for the oxygen analogue furan. The reason for this difference is the higher electronegativity for oxygen drawing away electrons to itself at the expense of the aromatic ring current. Yet as an aromatic substituent the thio group is less electron-releasing than the alkoxy group. Dibenzothiophenes, tricyclic heterocycles consisting of two benzene rings fused to a central thiophene ring, occurs widely in heavier fractions of petroleum.

## Thiols, Disulfides and Polysulfides

Thiol group contain the functionality R–SH. Thiols are structurally similar to the alcohol group, but these functionalities are very different in their chemical properties. Thiols are more nucleophilic, more acidic, and more readily oxidized. This acidity can differ by 5 p$K_a$ units.

The difference in electronegativity between sulfur (2.58) and hydrogen (2.20) is small and therefore hydrogen bonding in thiols is not prominent. Aliphatic thiols form monolayers on gold, which are topical in nanotechnology.

Certain aromatic thiols can be accessed through a Herz reaction.

Disulfides R–S–S–R with a covalent sulfur to sulfur bond are important for crosslinking: in biochemistry for the folding and stability of some proteins and in polymer chemistry for the crosslinking of rubber.

Longer sulfur chains are also known, such as in the natural product varacin which contains an unusual pentathiepin ring (5-sulfur chain cyclised onto a benzene ring).

## Thioesters

Thioesters have general structure R–C(O)–S–R. They are related to regular esters (R–C(O)–O–R) but are more susceptible to hydrolysis and related reactions. Thioesters are prominent in biochemistry, especially in fatty acid synthesis.

## Sulfoxides, Sulfones and Thiosulfinates

A sulfoxide, R–S(O)–R, is the *S*-oxide of a sulfide ("sulfide oxide"), a sulfone, R–S(O)$_2$–R, is the *S,S*-dioxide of a sulfide, a thiosulfinate, R–S(O)–S–R, is the *S*-oxide of a disulfide, and a thiosulfonate, R–S(O)$_2$–S–R, is the *S,S*-dioxide of a disulfide. All of these compounds are well known with extensive chemistry, e.g., dimethyl sulfoxide, dimethyl sulfone, and allicin.

## Sulfimides, Sulfoximides and Sulfonediimines

Sulfimides (also called a sulfilimines) are sulfur–nitrogen compounds of structure R$_2$S=NR′, the nitrogen analog of sulfoxides. They are of interest in part due to their pharmacological properties. When two different R groups are attached to sulfur, sulfimides are chiral. Sulfimides form stable α-carbanions.

Sulfoximides (also called sulfoximines) are tetracoordinate sulfur–nitrogen compounds, isoelectronic with sulfones, in which one oxygen atom of the sulfone is replaced by a substituted nitrogen atom, e.g., R$_2$S(O)=NR′. When two different R groups are attached to sulfur, sulfoximides are chiral. Much of the interest in this class of compounds is derived from the discovery that methionine sulfoximide (methionine sulfoximine) is an inhibitor of glutamine synthetase.

Sulfonediimines (also called sulfodiimines, sulfodiimides or sulfonediimides) are tetracoordinate sulfur–nitrogen compounds, isoelectronic with sulfones, in which both oxygen atoms of the sulfone are replaced by a substituted nitrogen atom, e.g., R$_2$S(=NR′)$_2$. They are of interest because of their biological activity and as building blocks for heterocycle synthesis.

## S-Nitrosothiols

*S*-Nitrosothiols, also known as thionitrites, are compounds containing a nitroso group attached to the sulfur atom of a thiol, e.g. R–S–N=O. They have received considerable attention in biochemistry because they serve as donors of the nitrosonium ion, $NO^+$, and nitric oxide, NO, which may serve as signaling molecules in living systems, especially related to vasodilation.

## Sulfur Halides

A wide range of organosulfur compounds are known which contain one or more halogen atom ("X" in the chemical formulas that follow) bonded to a single sulfur atom, e.g.: sulfenyl halides, RSX; sulfinyl halides, RS(O)X; sulfonyl halides, $RSO_2X$; alkyl and arylsulfur trichlorides, $RSCl_3$ and trifluorides, $RSF_3$; and alkyl and arylsulfur pentafluorides, $RSF_5$. Less well known are dialkylsulfur tetrahalides, mainly represented by the tetrafluorides, e.g., $R_2SF_4$.

## Thioketones, Thioaldehydes and Related Compounds

Compounds with double bonds between carbon and sulfur are relatively uncommon, but include the important compounds carbon disulfide, carbonyl sulfide, and thiophosgene. Thioketones (RC(=S)R′) are uncommon with alkyl substituents, but one example is thiobenzophenone. Thioaldehydes are rarer still, reflecting their lack of steric protection ("thioformaldehyde" exists as a cyclic trimer). Thioamides, with the formula $R_1C(=S)N(R_2)R_3$ are more common. They are typically prepared by the reaction of amides with Lawesson's reagent. Isothiocyanates, with formula R–N=C=S, are found naturally. Vegetable foods with characteristic flavors due to isothiocyanates include wasabi, horseradish, mustard, radish, Brussels sprouts, watercress, nasturtiums, and capers.

## S-oxides and S,S-dioxides of Thiocarbonyl Compounds

The *S*-oxides of thiocarbonyl compounds are known as thiocarbonyl *S*-oxides or sulfines, $R_2C=S=O$, and thiocarbonyl *S,S*-dioxides or sulfenes, $R_2C=SO_2$. These compounds are well known with extensive chemistry.

## Triple Bonds between Carbon and Sulfur

Triple bonds between sulfur and carbon in sulfaalkynes are rare and can be found in carbon monosulfide (CS) and have been suggested for the compounds $F_3CCSF_3$ and $F_5SCSF_3$. The compound HCSOH is also represented as having a formal triple bond.

## Thiocarboxylic Acids and Thioamides

Thiocarboxylic acids (RC(O)SH) and dithiocarboxylic acids (RC(S)SH) are well known. They are structurally similar to carboxylic acids but more acidic. Thioamides are analogous to amides.

## Sulfonic, Sulfinic and Sulfenic Acids, Esters, Amides and Related Compounds

Sulfonic acids have functionality $R–S(=O)_2–OH$. They are strong acids that are typically soluble in organic solvents. Sulfonic acids like trifluoromethanesulfonic acid is a frequently used reagent in organic chemistry. Sulfinic acids have functionality R–S(O)–H while sulfenic acids have functionality R–S–OH. In the series sulfonic—sulfinic—sulfenic acids, both the acid strength and stability diminish in that order. Sulfonamides, sulfinamides and sulfenamides, with formulas $R–SO_2NR'_2$, $R–S(O)NR'_2$, and $R–SNR'_2$, respectively, each have a rich chemistry. For example, sulfa drugs are sulfonamides derived from aromatic sulfonation. Chiral sulfinamides are used in asymmetric synthesis, while sulfenamides are used extensively in the vulcanization process to assist cross-linking. Thiocyanates, R–S–CN, are related to sulfenyl halides and esters in terms of reactivity.

## Sulfonium, Oxosulfonium and Related Salts

A sulfonium ion is a positively charged ion featuring three organic substituents attached to sulfur, with the formula $[R_3S]^+$. Together with their negatively charged counterpart, the anion, the compounds are called sulfonium salts. An oxosulfonium ion is a positively charged ion featuring three organic substituents and an oxygen attached to sulfur, with the formula $[R_3S{=}O]^+$. Together with their negatively charged counterpart, the anion, the compounds are called oxosulfonium salts. Related species include alkoxysulfonium and chlorosulfonium ions, $[R_2SOR]^+$ and $[R_2SCl]^+$, respectively.

## Sulfonium, Oxosulfonium and Thiocarbonyl Ylides

Deprotonation of sulfonium and oxosulfonium salts affords ylides, of structure $R_2S^+–C^-–R'_2$ and $R_2S(O)^+–C^-–R'_2$. While sulfonium ylides, for instance in the Johnson–Corey–Chaykovsky reaction used to synthesize oxiranes, are sometimes drawn with a C=S double bond, e.g., $R_2S{=}CR'_2$, the ylidic carbon–sulfur bond is highly polarized and is better described as being ionic. Sulfonium ylides are key intermediates in the synthetically useful Stevens rearrangement. Thiocarbonyl ylides ($RR'C{=}S^+–C^-–RR'$) can form by ring-opening of thiiranes, photocyclization of aryl vinyl sulfides, as well as by other processes.

## Sulfuranes and Persulfuranes

Sulfuranes are relatively specialized functional group that are tetravalent, hypervalent sulfur compounds, with the formula $SR_4$ and likewise persulfuranes are hexavalent $SR_6$. All-carbon hexavalent complexes have been known for the heavier representatives of the chalcogen group, for instance the compound hexamethylpertellurane ($Te(Me)_6$) was discovered in 1990 by reaction of tetramethyltellurium with xenon difluoride to $TeF_2(CH_3)_4$ followed by reaction with dimethylzinc. The sulfur analogue hexamethylpersulfurane ($S(CH_3)_6$) has been predicted to be stable but has not been synthesized yet.

The first ever all-carbon persulfurane actually synthesized in a laboratory has two methyl and two biphenyl ligands:

It is prepared from the corresponding sulfurane 1 with xenon difluoride/boron trifluoride in acetonitrile to the sulfuranyl dication 2 followed by reaction with methyllithium in tetrahydrofuran to (a stable) persulfurane 3 as the cis isomer. X-ray diffraction shows C–S bond lengths ranging between 189 and 193 pm (longer than the standard bond length) with the central sulfur atom in a distorted octahedral molecular geometry.

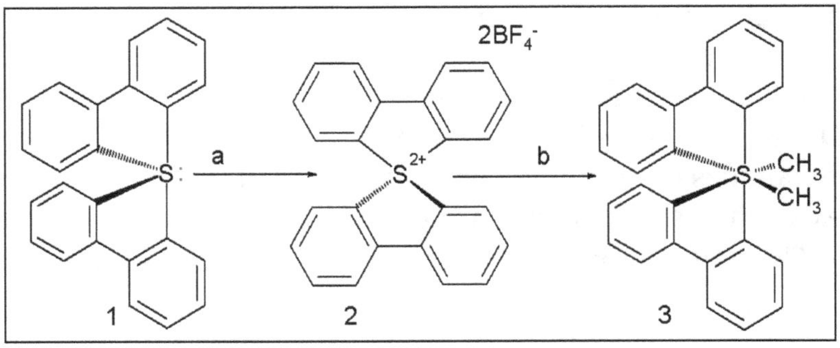


Computer simulation suggests that these bonds are very polar with the negative charges residing on carbon.

### Naturally Occurring Organosulfur Compounds

Not all organosulfur compounds are foul-smelling pollutants. Penicillin and cephalosporin are life-saving antibiotics, derived from fungi. Gliotoxin is a sulfur-containing mycotoxin produced by several species of fungi under investigation as an antiviral agent. Compounds like allicin and ajoene are responsible for the odor of garlic, and lenthionine contributes to the flavor of shiitake mushrooms. Volatile organosulfur compounds also contribute subtle flavor characteristics to wine, nuts, cheddar cheese, chocolate, coffee, and tropical fruit flavors. Many of these natural products also have important medicinal properties such as preventing platelet aggregation or fighting cancer.

## Acetylacetone

Acetylacetone is an organic compound with the formula $CH_3COCH_2COCH_3$. It is a colorless liquid, classified as a 1,3-diketone. It exists in equilibrium with a tautomer $CH_3C(O)CH=C(OH)CH_3$. These tautomers interconvert so rapidly under most conditions that they are treated as a single compound in most applications. It is a colorless liquid that is a precursor to acetylacetonate anion (commonly abbreviated acac$^-$), a bidentate ligand. It is also a building block for the synthesis of heterocyclic compounds.

### Properties

### Tautomerism

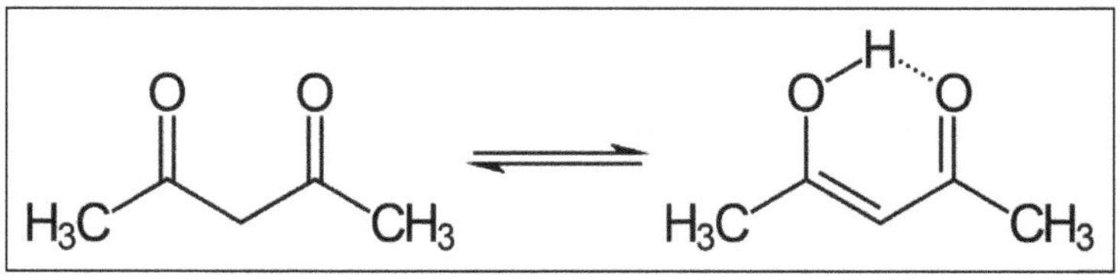


The keto and enol tautomers of acetylacetone coexist in solution. The enol form has $C_{2v}$ symmetry, meaning the hydrogen atom is shared equally between the two oxygen atoms. In the gas phase, the equilibrium constant, $K_{keto\rightarrow enol}$, is 11.7, favoring the enol form. The two tautomeric forms can be distinguished by NMR spectroscopy, IR spectroscopy and other methods.

| Solvent | $K_{keto \rightarrow enol}$ |
|---|---|
| Gas phase | 11.7 |
| Cyclohexane | 42 |
| Toluene | 10 |
| THF | 7.2 |
| DMSO | 2 |
| Water | 0.23 |

The equilibrium constant tends to be high in nonpolar solvents; the keto form becomes more favorable in polar, hydrogen-bonding solvents, such as water. The enol form is a vinylogous analogue of a carboxylic acid.

## Acid–base Properties

| Solvent | T/°C | $pK_a$ |
|---|---|---|
| 40% ethanol/water | 30 | 9.8 |
| 70% dioxane/water | 28 | 12.5 |
| 80% DMSO/water | 25 | 10.16 |
| DMSO | 25 | 13.41 |

Acetylacetone is a weak acid:

$$C_5H_8O_2 \rightleftharpoons C_5H_7O_2^- + H^+$$

IUPAC recommended p*Ka* values for this equilibrium in aqueous solution at 25 °C are 8.99 ± 0.04 ($I$ = 0), 8.83 ± 0.02 ($I$ = 0.1 M $NaClO_4$) and 9.00 ± 0.03 ($I$ = 1.0 M $NaClO_4$; $I$ = Ionic strength). Values for mixed solvents are available. Very strong bases, such as organolithium compounds, will deprotonate acetylacetone twice. The resulting dilithio species can then be alkylated at C-1.

## Preparation

Acetylacetone is prepared industrially by the thermal rearrangement of isopropenyl acetate.

$$CH_2(CH_3)COC(O)Me \rightarrow MeC(O)CH_2C(O)Me$$

Laboratory routes to acetylacetone begin also with acetone. Acetone and acetic anhydride upon the addition of boron trifluoride ($BF_3$) catalyst:

$$(CH_3CO)_2O + CH_3C(O)CH_3 \rightarrow CH_3C(O)CH_2C(O)CH_3$$

A second synthesis involves the base-catalyzed condensation of acetone and ethyl acetate, followed by acidification:

$$NaOEt + EtO_2CCH_3 + CH_3C(O)CH_3 \rightarrow NaCH_3C(O)CHC(O)CH_3 + 2\ EtOH$$

$$NaCH_3C(O)CHC(O)CH_3 + HCl \rightarrow CH_3C(O)CH_2C(O)CH_3 + NaCl$$

Because of the ease of these syntheses, many analogues of acetylacetonates are known. Some examples include $C_6H_5C(O)CH_2C(O)C_6H_5$ (dbaH) and $(CH_3)_3CC(O)CH_2C(O)CC(CH_3)_3$. Hexafluoroacetylacetonate is also widely used to generate volatile metal complexes.

## Reactions

### Condensations

Acetylacetone is a versatile bifunctional precursor to heterocycles because both keto groups undergo condensation. Hydrazine reacts to produce pyrazoles. Urea gives pyrimidines. Condensation with two aryl- and alkylamines to gives NacNacs, wherein the oxygen atoms in acetylacetone are replaced by NR (R = aryl, alkyl).

### Coordination Chemistry

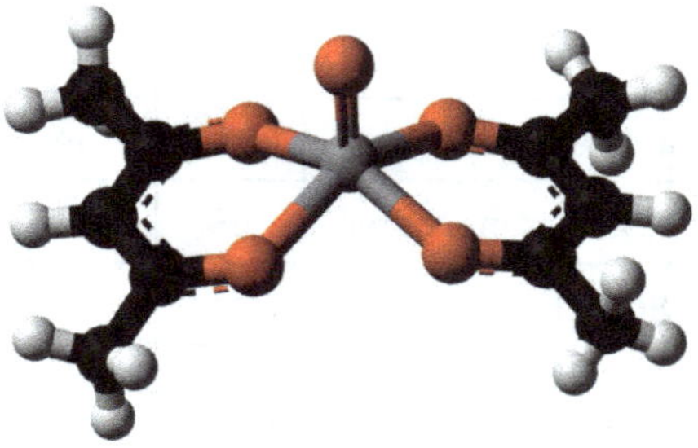

A ball-and-stick model of $VO(acac)_2$

The acetylacetonate anion, $acac^-$, forms complexes with many transition metal ions. A general method of synthesis is to treat a metal salt with acetylacetone in the presence of a base:

$$MB_z + z\ Hacac \rightleftharpoons M(acac)_z + z\ BH$$

Both oxygen atoms bind to the metal to form a six-membered chelate ring. In some cases the chelate effect is so strong that no added base is needed to form the complex.

### Biodegradation

Enzymatic breakdown: The enzyme acetylacetone dioxygenase cleaves the carbon-carbon bond of acetylacetone, producing acetate and 2-oxopropanal. The enzyme is iron(II)-dependent, but it has been proven to bind to zinc as well. Acetylacetone degradation has been characterized in the bacterium *Acinetobacter johnsonii*.

$$C_5H_8O_2 + O_2 \rightarrow C_2H_4O_2 + C_3H_4O_2$$

# Organophosphorus Compound

Organophosphorus compounds are organic compounds containing phosphorus. They are used primarily in pest control as an alternative to chlorinated hydrocarbons that persist in the environment.

Some organophosphorus compounds are highly effective insecticides, although some are extremely toxic to man, including sarin and VX nerve agents.

Organophosphorus chemistry is the corresponding science of the properties and reactivity of organophosphorus compounds. Phosphorus, like nitrogen, is in group 15 of the periodic table, and thus phosphorus compounds and nitrogen compounds have many similar properties. The definition of organophosphorus compounds is variable, which can lead to confusion. In industrial and environmental chemistry, an organophosphorus compound need contain only an organic substituent, but need not have a direct phosphorus-carbon (P-C) bond. Thus a large proportion of pesticides (e.g., malathion), are often included in this class of compounds.

Phosphorus can adopt a variety of oxidation states, and it is general to classify organophosphorus compounds based on their being derivatives of phosphorus (V) vs phosphorus (III), which are the predominant classes of compounds. In a descriptive but only intermittently used nomenclature, phosphorus compounds are identified by their coordination number σ and their valency λ. In this system, a phosphine is a $\sigma^3\lambda^3$ compound.

## Organophosphorus (V) Compounds

### Phosphate Esters and Amides

Phosphate esters have the general structure $P(=O)(OR)_3$ feature P(V). Such species are of technological importance as flame retardant agents, and plasticizers. Lacking a P–C bond, these compounds are in the technical sense not organophosphorus compounds but esters of phosphoric acid. Many derivatives are found in nature, such as phosphatidylcholine. Phosphate ester are synthesized by alcoholysis of phosphorus oxychloride. A variety of mixed amido-alkoxo derivatives are known, one medically significant example being the anti-cancer drug cyclophosphamide. Also derivatives containing the thiophosphoryl group (P=S) include the pesticide malathion. The organophosphates prepared on the largest scale are the zinc dithiophosphates, as additives for motor oil. Several million kilograms of this coordination complex are produced annually by the reaction of phosphorus pentasulfide with alcohols.

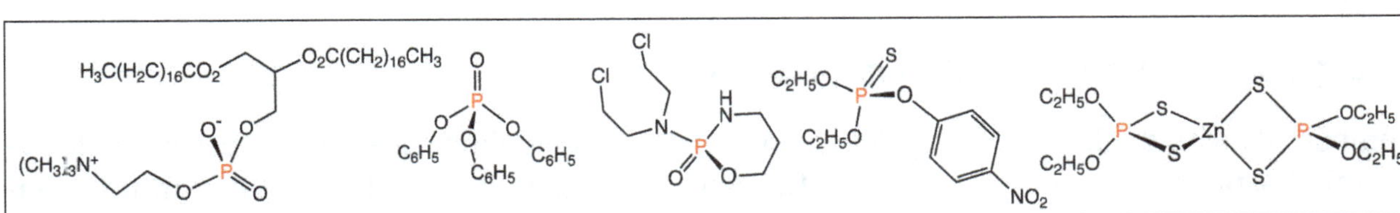

Illustrative organophosphates and related compounds: phosphatidylcholine, triphenylphosphate, cyclophosphamide, parathion, and zinc dithiophosphate.

In the environment, these compounds break down via hydrolysis to eventually afford phosphate and the organic alcohol or amine from which they are derived.

### Phosphonic and Phosphinic Acids and their Esters

Phosphonates are esters of phosphonic acid and have the general formula $RP(=O)(OR')_2$. Phosphonates have many technical applications, a well-known member being glyphosate, better known as Roundup. With the formula $(HO)_2P(O)CH_2NHCH_2CO_2H$, this derivative of glycine is one of the most widely used herbicides. Bisphosphonates are a class of drugs to treat osteoporosis. The nerve gas agent sarin, containing both C–P and F–P bonds, is a phosphonate.

Phosphinates feature *two* P–C bonds, with the general formula $R_2P(=O)(OR')$. A commercially significant member is the herbicide Glufosinate, it has the structure $CH_3P(O)(OH)CH_2CH_2CH(NH_2)CO_2H$.

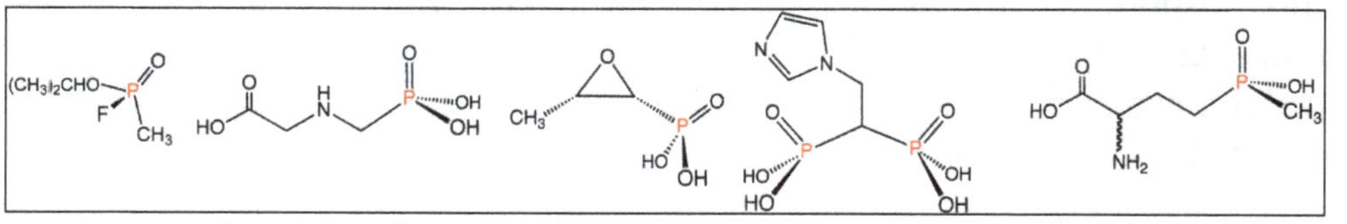

Illustrative examples of phosphonates and phosphinates in the order shown: Sarin (phosphonate), Glyphosate (phosphonate), fosfomycin (phosphonate), zoledronic acid (phosphonate), and Glufosinate (phosphinate). In aqueous solution, phosphonic acids ionize to give the corresponding organophosphonates.

The Michaelis–Arbuzov reaction is the main method for the synthesis of these compounds. For example, dimethylmethylphosphonate arises from the rearrangement of trimethylphosphite, which is catalyzed by methyl iodide. In the Horner–Wadsworth–Emmons reaction and the Seyferth–Gilbert homologation, phosphonates are used in reactions with carbonyl compounds. The Kabachnik–Fields reaction is a method for the preparation of aminophosphonates. These compounds contain a very inert bond between phosphorus and carbon. Consequently, they hydrolyze to give phosphonic and phosphinic acid derivatives, but not phosphate.

## Phosphine Oxides, Imides and Chalcogenides

Phosphine oxides (designation $\sigma^4\lambda^5$) have the general structure $R_3P=O$ with formal oxidation state V. Phosphine oxides form hydrogen bonds and some are therefore soluble in water. The P=O bond is very polar with a dipole moment of 4.51 D for triphenylphosphine oxide.

Compounds related to phosphine oxides include phosphine imides ($R_3PNR'$) and related chalcogenides ($R_3PE$, where E = S, Se, Te). These compounds are some of the most thermally stable organophosphorus compounds.

## Phosphonium Salts and Phosphoranes

Compounds with the formula $[PR_4^+]X^-$ comprise the phosphonium salts. These species are tetrahedral phosphorus(V) compounds. From the commercial perspective, the most important member is tetrakis(hydroxymethyl)phosphonium chloride, $[P(CH_2OH)_4]Cl$, which is used as a fire retardant in textiles. Approximately 2M kg are produced annually of the chloride and the related sulfate. They are generated by the reaction of phosphine with formaldehyde in the presence of the mineral acid:

$$PH_3 + HX + 4\,CH_2O \rightarrow [P(CH_2OH)_4^+]X^-$$

A variety of phosphonium salts can be prepared by alkylation and arylation of organophosphines:

$$PR_3 + R'X \rightarrow [PR_3R'^+]X^-$$

The methylation of triphenylphosphine is the first step in the preparation of the Wittig reagent.

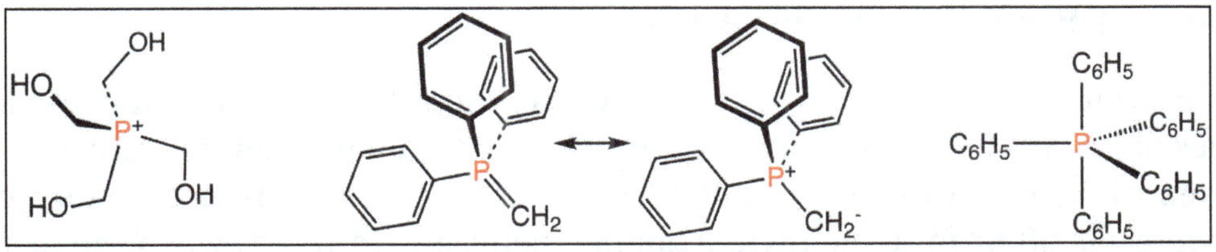


Illustrative phosphorus(V) compounds: the phosphonium ion $P(CH_2OH)_4^+$, two resonance structures for the Wittig reagent $Ph_3PCH_2$, and pentaphenylphosphorane, a rare pentaorganophophorus compound.

The parent phosphorane ($\sigma^5\lambda^5$) is $PH_5$, which is unknown. Related compounds containing both halide and organic substituents on phosphorus are fairly common. Those with five organic substituents are rare, although $P(C_6H_5)_5$ is known, being derived from $P(C_6H_5)_4^+$ by reaction with phenyllithium.

Phosphorus ylides are unsaturated phosphoranes, known as Wittig reagents, e.g. $CH_2P(C_6H_5)_3$. These compounds feature tetrahedral phosphorus(V) and are considered relatives of phosphine oxides. They also are derived from phosphonium salts, but by deprotonation not alkylation.

## Organophosphorus (III) Compounds

### Phosphites, Phosphonites and Phosphinites

Phosphites, sometimes called phosphite esters, have the general structure $P(OR)_3$ with oxidation state +3. Such species arise from the alcoholysis of phosphorus trichloride:

$$PCl_3 + 3\ ROH \rightarrow P(OR)_3 + 3\ HCl$$

The reaction is general, thus a vast number of such species are known. Phosphites are employed in the Perkow reaction and the Michaelis–Arbuzov reaction. They also serve as ligands in organometallic chemistry.

Intermediate between phosphites and phosphines are phosphonites ($P(OR)_2R'$) and phosphinite ($P(OR)R'_2$). Such species arise via alcoholysis reactions of the corresponding phosphinous and phosphonous chlorides (($PClR'_2$) and $PCl_2R'$, respectively).

### Phosphines

The parent compound of the phosphines is $PH_3$, called phosphine in the US and British Commonwealth, but phosphane elsewhere. Replacement of one or more hydrogen centers by an organic substituents (alkyl, aryl), gives $PH_{3-x}R_x$, an organophosphine, generally referred to as phosphines.

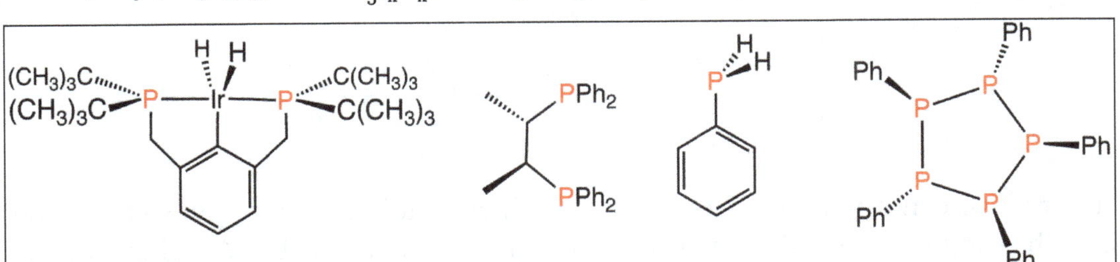


Various reduced organophosphorus compounds: a complex of an organophosphine pincer ligand, the chiral diphosphine used in homogeneous catalysis, the primary phosphine $PhPH_2$, and the phosphorus(I) compound $(PPh)_5$.

## Comparison of Phosphines and Amines

The phosphorus atom in phosphines has a formal oxidation state −3 ($\sigma^3\lambda^3$) and are the phosphorus analogues of amines. Like amines, phosphines have a trigonal pyramidal molecular geometry although often with smaller C-E-C angles (E = N, P), at least in the absence of steric effects. The C-P-C bond angle is 98.6° for trimethylphosphine increasing to 109.7° when the methyl groups are replaced by *tert*-butyl groups. When used as ligands, the steric bulk of tertiary phosphines is evaluated by their cone angle. The barrier to pyramidal inversion is also much higher than nitrogen inversion to occur, and therefore phosphines with three different substituents can be resolved into thermally stable optical isomers. Phosphines are often less basic than corresponding amines, for instance the phosphonium ion itself has a p*K*a of −14 compared to 9.21 for the ammonium ion; trimethylphosphonium has a p*K*a of 8.65 compared to 9.76 for trimethylammonium. However, triphenylphosphine ($pK_a$ 2.73) is more basic than triphenylamine ($pK_a$ −5), mainly because the lone pair of the nitrogen in $NPh_3$ is partially delocalized into the three phenyl rings. Whereas the lone pair on nitrogen is delocalized in pyrrole, the lone pair on phosphorus atom in the phosphorus equivalent of pyrrole (phosphole) is not. The reactivity of phosphines matches that of amines with regard to nucleophilicity in the formation of phosphonium salts with the general structure $PR_4^+X^-$. This property is used in the Appel reaction for converting alcohols to alkyl halides. Phosphines are easily oxidized to the corresponding phosphine oxides, whereas amine oxides are less readily generated. In part for this reason, phosphines are very rarely encountered in nature.

## Synthetic Routes

From the commercial perspective, the most important phosphine is triphenylphosphine, several million kilograms being produced annually. It is prepared from the reaction of chlorobenzene, PCl3, and sodium. Phosphines of a more specialized nature are usually prepared by other routes. Phosphorus halides undergo Nucleophilic displacement by organometallic reagents such as Grignard reagents. Conversely, some syntheses entail nucleophilic displacement of phosphide anion equivalents ("$R_2P^-$") by aryl- and alkyl halides. Primary ($RPH_2$) and secondary phosphines (RRPH and $R_2PH$) add to alkenes in presence of a strong base (e.g., KOH in DMSO). Markovnikov's rules apply. Similar reactions occur involving alkynes. Base is not required for electron-deficient alkenes (e.g., derivatives of acrylonitrile) and alkynes.

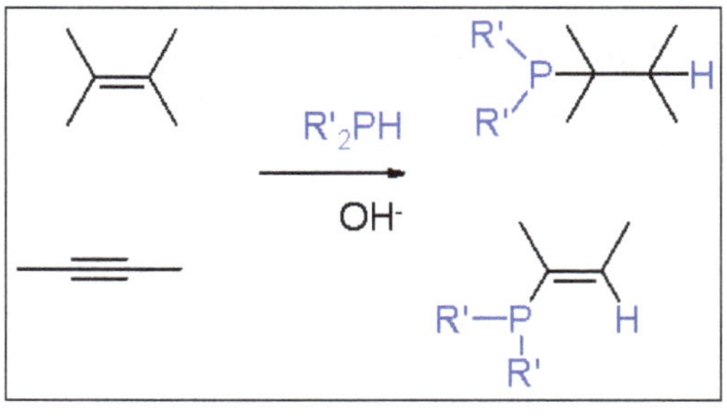


Under free-radical conditions the P-H bonds of primary and secondary phosphines add across alkenes. Such reactions proceed with anti-Markovnikov regiochemistry. AIBN or organic peroxides are used as initiators. Tertiary phosphine oxides and sulfides can be reduced with chlorosilanes and other reagents.

## Reactions

Organophosphines are nucleophiles and ligands. Two major applications are as reagents in the Wittig reaction and as supporting phosphine ligands in homogeneous catalysis.

Their nucleophilicity is evidenced by their reactions with alkyl halides to give phosphonium salts. Phosphines are nucleophilic catalysts in organic synthesis, e.g. the Rauhut–Currier reaction and Baylis-Hillman reaction.

Phosphines are reducing agents, as illustrated in the Staudinger reduction for the conversion of organic azides to amines and in the Mitsunobu reaction for converting alcohols into esters. In these processes, the phosphine is oxidized to phosphorus (V). Phosphines have also been found to reduce activated carbonyl groups, for instance the reduction of an α-keto ester to an α-hydroxy ester. In the proposed reaction mechanism, the first proton is on loan from the methyl group in trimethylphosphine (triphenylphosphine does not react).

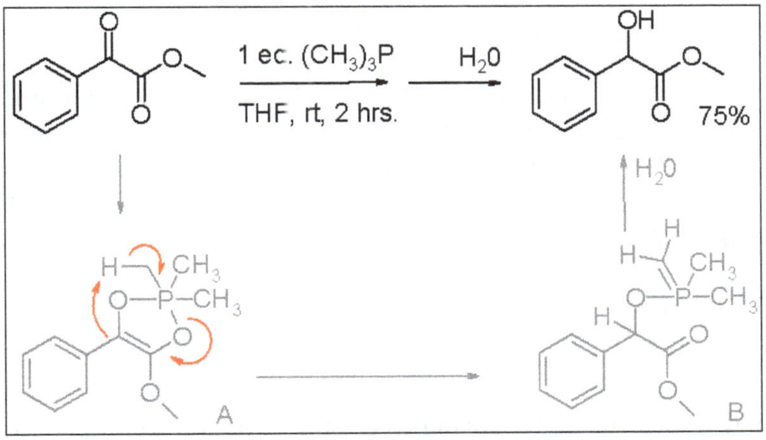


Reduction of activated carbonyl groups by alkyl phosphines.

## Primary and Secondary Phosphines

In addition to the other reactions associated with phosphines, those bearing P-H groups exhibit additional reactivity associated with the P-H bonds. They are readily deprotonated using strong bases to give phosphide anions. Primary and secondary phosphines are generally prepared by reduction of related phosphorus halides or esters. For example, phosphonates are reduced to primary phosphines:

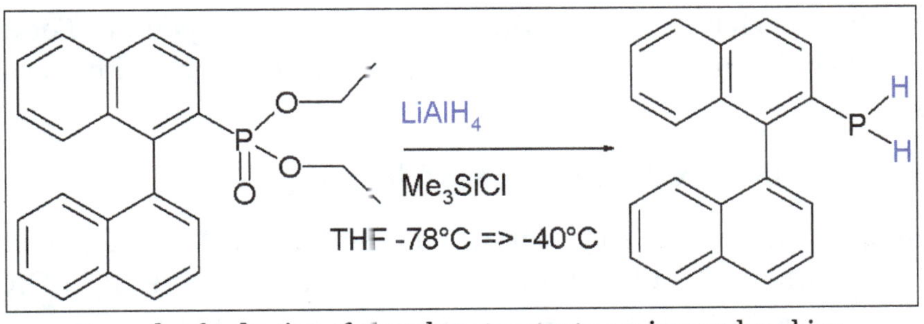


Example of reduction of phosphonate ester to a primary phosphine.

The stability is attributed to conjugation between the aromatic ring and the phosphorus lone pair.

### Phosphaalkenes and Phosphaalkynes

Compounds with carbon phosphorus (III) multiple bonds are called phosphaalkenes ($R_2C=PR$) and phosphaalkynes ($RC\equiv P$). They are similar in structure, but not in reactivity, to imines ($R_2C=NR$) and nitriles ($RC\equiv N$), respectively. In the compound phosphorine, one carbon atom in benzene is replaced by phosphorus. Species of this type are relatively rare but for that reason are of interest to researchers. A general method for the synthesis of phosphaalkenes is by 1,2-elimination of suitable precursors, initiated thermally or by base such as DBU, DABCO, or triethylamine:

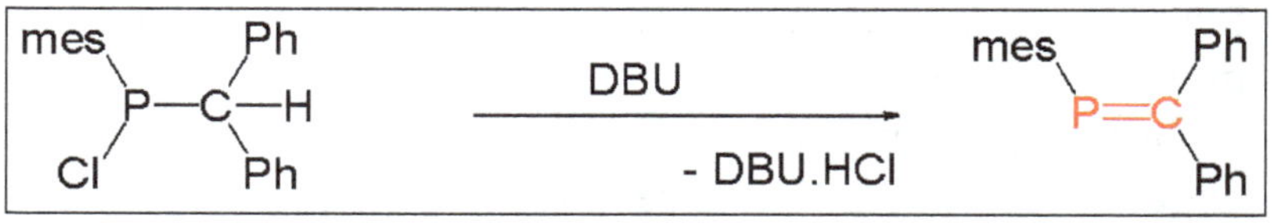


Thermolysis of $Me_2PH$ generates $CH_2=PMe$, an unstable species in the condensed phase.

### Organophosphorus (0), (I) and (II) Compounds

Compounds where phosphorus exists in a formal oxidation state of less than III are uncommon, but examples are known for each class. Organophosphorus(0) species are debatably illustrated by the carbene adducts, $[P(NHC)]_2$, where NHC is an N-heterocyclic carbene. With the formulae $(RP)_n$ and $(R_2P)_2$, respectively, compounds of phosphorus(I) and (II) are generated by reduction of the related organophosphorus(III) chlorides:

$$5\ PhPCl_2 + 5\ Mg \rightarrow (PhP)_5 + 5\ MgCl_2$$

$$2\ Ph_2PCl + Mg \rightarrow Ph_2P\text{-}PPh_2 + MgCl_2$$

Diphosphenes, with the formula $R_2P_2$, formally contain phosphorus-phosphorus double bonds. These phosphorus(I) species are rare but are stable provided that the organic substituents are large enough to prevent catenation. Many mixed-valence compounds are known, e.g. the cage $P_7(CH_3)_3$.

## Organosilicon Compound

Organosilicon compound referred to as "organic silicon" is the organic compound containing carbon - silicon bond inside the molecule structure. It has similarity in its structure with the organic carbon compounds but also has difference with organic carbon compounds. For example, the silicone molecule silicon - only contains single bond between silicon atoms without containing double and triple bonds.

The organosilicon compounds are divided into the following groups:

1. The organohalosilanes, or alkyl (aryl)-halosilanes, $R_nSiX_4\text{-}n$; organohydrohalosilanes $R_nSiH_mX_{4-(n+m)}$, where X is Cl in most cases; organoalkoxysilanes $R_nSi(OR')_{4-n}$; organoacyloxysilanes $R_nSi(OCOR')_{4-n}$; and organoaminosilanes $R_nSi(NH_2)_{4-n}$.
2. The organosilanes (substituted silanes) $R_nSiH_{4-n}$.

3. The organosiloxanes, including disiloxanes $R_3SiOSiR_3$, trisiloxanes $R_3SiOSi(R_2)OSiR_3$, and higher siloxanes; cyclosiloxanes $(R_2SiO)n$, where $n$ = 3-10 (most often 3-4).

4. Heterocyclic compounds, such as dimethylsilacyclobutane:

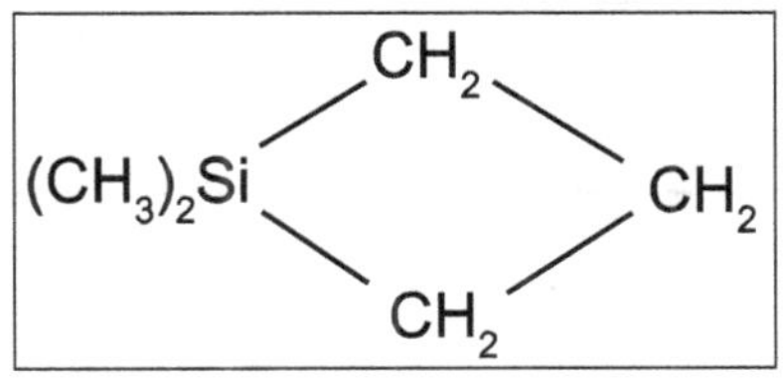


Compounds of the first two groups resemble the corresponding chemically analogous inorganic compounds, such as the halosilanes $SiH_mX_{4-n}$, silanes $Si_xH_y$, and alkoxysilanes $Si(OR)_4$.

*Preparation.* The most important organosilicon compounds are produced on an industrial scale mainly by direct reaction of alkyl (aryl) chlorides with silicon in the presence of copper. In addition to the principal product of the reaction $2RCl + Si \rightarrow R_2SiCl_2$, a mixture of various by products is also formed ($RSiCl_3$, $R_3SiCl$, $RHSiCl_2$, and so on); these by products also have industrial uses, such as in the synthesis of organosilicon compounds with more complicated structures, or polymers. Organosilicon compounds are also produced by reactions with organometallic compounds, $CH_3SiCl_3 + C_6H_5MgCl \rightarrow CH_3(C_6H_5)SiCl_2 + MgCl_2$; by thermal condensation, $CH_3SiHCl_2 + CH_2 = CHCl \rightarrow CH_3(CH_2 = CH)SiCl_2 + HCl$; by liquid-phase dehydrocondensation, $CH_3SiHCl_2 + C_6H_6 \rightarrow CH_3(C_6H_5)SiCl_2 + H_2$; and by the addition of organohydrochlorosilanes to unsaturated compounds, $CH_3SiHCl_2 + RCH{=}CH_2 \rightarrow CH_3(RCH_2CH_2)SiCl_2$. Alkoxysilanes and acyloxysilanes are most frequently prepared by the reaction of organochlorosilanes with alcohols, acids, and other reagents.

*Properties and use.* Almost all organosilicon compounds are colorless liquids; only some of them, such as the cyclosiloxanes $(R_2SiO)_3$, are solid crystalline substances. Organosilicon compounds are usually thermally stable (~600°C), mabe distilled at atmospheric pressure and under vacuum, are highly soluble in hydrocarbons, chlorinated hydrocarbons, ethers, and other organic solvents, and are immiscible with water. Organosilicon compounds are readily hydrolyzed, especially the organochlorosilanes, which fume in air:

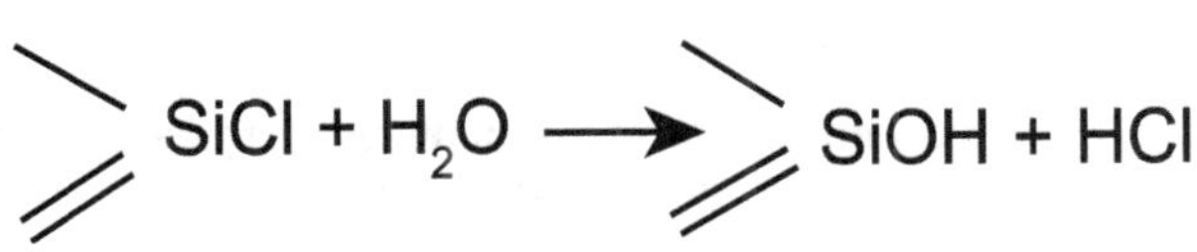


The organohydroxysilanes that form during hydrolysis undergo intermolecular condensations leading to the formation of organosiloxanes:

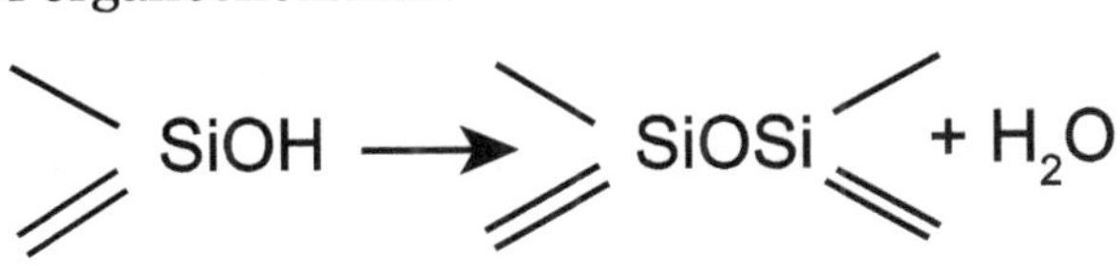


The evolved or added acid causes polycondensation, with the formation of silicones. Depending on the number of hydrolyzable groups (usually chlorine; sometimes alkoxy, acyloxy, or amino groups), a distinction is made among mono-, di-, tri-, and tetrafunctional organosilicon compounds, which hydrolyze to disiloxanes, mixtures of linear polymers $HO(R_2SiO)_nH$ and low-molecular weight cyclic diorganosiloxanes $(R_2SiO)_n$ (usually, $n$ = 4), and polymers of cyclolinear and crosslinked structure $(RSiO_{1.5})_n$ and $(SiO_2)_n$, respectively.

| Table: Properties of some organosilicon compounds | | |
|---|---|---|
| | Boiling point (°C) | Density at 20 °C (g/cm³) |
| Dimethylclichlorasilane $(CH_3)_2SiCI_2$ | 70.1 | 1.0637 |
| Trimethylchlorosilane $(CH_3)_3SiCI$ | 57.7 | 0.8580 |
| Diethyldichlorosilane $(C_2H_8)_2SiCI_2$ | 128.9 | 1.0504 |
| Triethylchlorosilane $(C_2H_5)_3SiCl$ | 146.0 | 0.8968 |
| Vinyltrichlorosilane $CH_2=CHSiCl_3$ | 92.5 | 1.2426 |
| Phenyltrichlorosilane $C_6H_5SiCl_3$ | 201.0 | 1.3240 |
| Diphenyldichlorosilane $(C_6H_5)_2SiCl_2$ | 304.0 | 1.2216 |
| Methylphenyldichlorosilane $CH_3(C_6H_5)SiCl_2$ | 205.5 | 1.1866 |
| Methyltrimethoxysilane $CH_3Si(OCH_3)_3$ | 103-105 | – |
| Methyltriethoxysilane $CH_3Si(OC_2H_3)_3$ | 151 | 0.9383 |
| Ethyltriethoxysilane $C_2H_5Si(OC_2H_5)_3$ | 159 | 0.9207 |
| Tetraethoxysilane $(C_2H_5O)_4Si$ | 166.5 | 0.9676 |

Cyclic diorganosiloxanes (mainly trimers and tetramers), which are also formed during the thermocatalytic degradation (alkali, 150° 400 °C) of polymers containing terminal hydroxyl groups, are used in the production of silicone rubbers and silicone oils. A mixture of tetraethoxysilane with the products of its partial hydrolysis is used under the name "ethyl silicate" for the preparation of molds for precision casting.

## References


- Silberberg, Martin (2004). Chemistry: The Molecular Nature Of Matter and Change. New York: McGraw-Hill Companies. ISBN 0-07-310169-9
- Aliphatic-hydrocarbons, chemistry: byjus.com, Retrieved 26 May, 2019
- Borak J.; Diller W. F. (2001). "Phosgene exposure: mechanisms of injury and treatment strategies". Journal of Occupational and Environmental Medicine. 43 (2): 110–9. doi:10.1097/00043764-200102000-00008. PMID 1122762
- Weast, Robert C., ed. (1981). CRC Handbook of Chemistry and Physics (62nd ed.). Boca Raton, FL: CRC Press. pp. C–301, E–61. ISBN 0-8493-0462-8
- Aromatic-compound, science: britannica.com, Retrieved 19 July, 2019
- Heterocyclic-compound, science: britannica.com, Retrieved 13 August, 2019
- Spence, Robert; Wild, William (1935). "114. The vapour-pressure curve of formaldehyde, and some related data". Journal of the Chemical Society (Resumed): 506–509. doi:10.1039/jr9350000506
- Organosilicon-Compounds: encyclopedia2.thefreedictionary.com, Retrieved 11 April, 2019

# 3
# Organic Reactions

The chemical reactions which involve organic compounds are known as organic reactions. Some of the major types of organic reactions are substitution reactions, addition reactions, elimination reactions, rearrangement reactions and oxidation-reduction reactions. The chapter closely examines these reactions to provide an extensive understanding of the subject.

Organic reactions are the chemical reactions that are undergone by organic compounds (the chemical compounds containing carbon). A few important types of organic reactions are illustrated below:

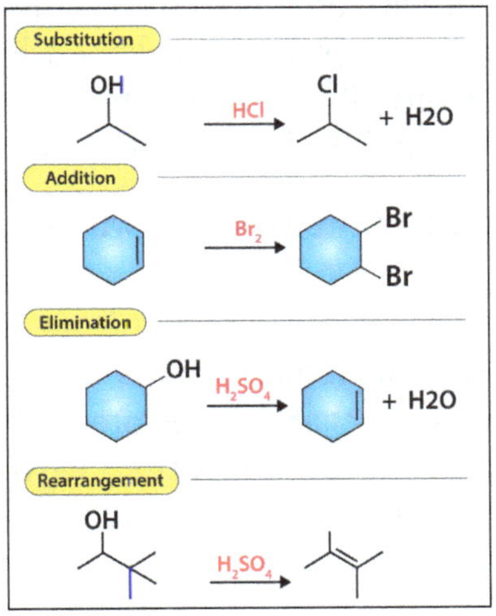


## Substitution Reaction

Substitution reaction (also known as single displacement reaction or single substitution reaction) is a chemical reaction during which one functional group in a chemical compound is replaced by another functional group. Substitution reactions are of prime importance in organic chemistry. Substitution reactions in organic chemistry are classified either as electrophilic or nucleophilic depending upon the reagent involved.

Organic substitution reactions are classified in several main organic reaction types depending on whether the reagent that brings about the substitution is considered an electrophile or a nucleophile, whether a reactive intermediate involved in the reaction is a carbocation, a carbanion or a free radical or whether the substrate is aliphatic or aromatic. Detailed understanding of a reaction type helps to predict the product outcome in a reaction. It also is helpful for optimizing a reaction with regard to variables such as temperature and choice of solvent.

A good example of a substitution reaction is halogenation. When chlorine gas ($Cl_2$) is irradiated, some of the molecules are split into two chlorine radicals (Cl•) whose free electrons are strongly nucleophilic. One of them breaks a C–H covalent bond in $CH_4$ and grabs the hydrogen atom to form the electrically neutral HCl. The other radical reforms a covalent bond with the $CH_3$• to form $CH_3Cl$ (methyl chloride).

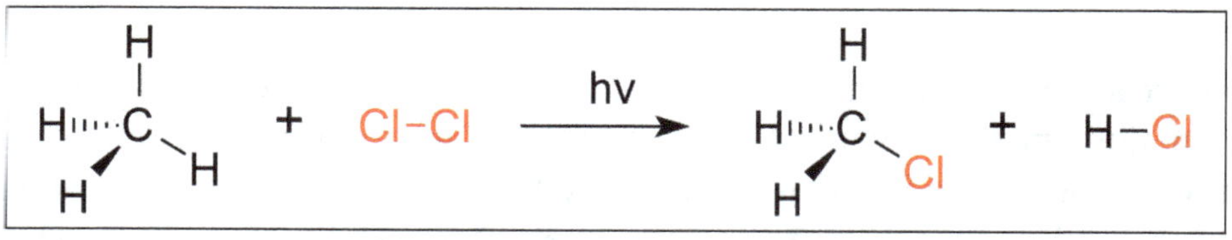


Chlorination of methane by chlorine.

## Nucleophilic Substitution

In organic (and inorganic) chemistry, nucleophilic substitution is a fundamental class of reactions in which a nucleophile selectively bonds with or attacks the positive or partially positive charge on an atom or a group of atoms. As it does so, it replaces a weaker nucleophile which then becomes a leaving group; The remaining positive or partially positive atom becomes an electrophile. The whole molecular entity of which the electrophile and the leaving group are part is usually called the substrate.

The most general form for the reaction may be given as where R-LG indicates the substrate.

Nuc: + R-LG → R-Nuc + LG:

The electron pair (:) from the nucleophile (Nuc:) attacks the substrate (R-LG) forming a new covalent bond Nuc-R-LG. The prior state of charge is restored when the leaving group (LG) departs with an electron pair. The principal product in this case is R-Nuc. In such reactions, the nucleophile is usually electrically neutral or negatively charged, whereas the substrate is typically neutral or positively charged.

An example of nucleophilic substitution is the hydrolysis of an alkyl bromide, R-Br, under basic conditions, where the *attacking* nucleophile is the base $OH^-$ and the leaving group is $Br^-$.

$$R\text{-}Br + OH^- \rightarrow R\text{-}OH + Br^-$$

Nucleophilic substitution reactions are commonplace in organic chemistry, and they can be broadly categorized as taking place at a carbon of a saturated aliphatic compound carbon or (less often) at an aromatic or other unsaturated carbon center.

## Mechanisms

These substitutions can be produced by two different mechanisms categorized at: unimolecular nucleophilic substitution ($S_N1$) and bimolecular nucleophilic substitution ($S_N2$).

The $S_N1$ mechanism has two steps. In the first step, the leaving group departs, forming a carbocation

$C^+$. In the second step, the nucleophilic reagent (Nuc:) attaches to the carbocation and forms a covalent sigma bond. If the substrate has a chiral carbon, this mechanism can result in either inversion of the stereochemistry or retention of configuration. Usually both occur without preference. The result is racemization.

The $S_N2$ mechanism has just one step. The attack of the reagent and the expulsion of the leaving group happen simultaneously. This mechanism always results in inversion of configuration.

If the substrate that is under nucleophilic attack is chiral, the reaction will lead to an inversion of its stereochemistry called a Walden inversion.

$S_N2$ attack may occur if the backside route of attack is not sterically hindered by substituents on the substrate. Therefore, this mechanism usually occurs at an unhindered primary carbon center. If there is steric crowding on the substrate near the leaving group, such as at a tertiary carbon center, the substitution will involve an $S_N1$ rather than an $S_N2$ mechanism, (an $S_N1$ would also be more likely in this case because a sufficiently stable carbocation intermediary could be formed).

When the substrate is an aromatic compound the reaction type is nucleophilic aromatic substitution. Carboxylic acid derivatives react with nucleophiles in nucleophilic acyl substitution. This kind of reaction can be useful in preparing compounds.

## Electrophilic Substitution

Electrophiles are involved in electrophilic substitution reactions, particularly in electrophilic aromatic substitutions.

In this example the benzene ring's electron resonance structure is attacked by an electrophile $E^+$. The resonating bond is broken and a carbocation resonating structure results. Finally a proton is kicked out and a new aromatic compound is formed.

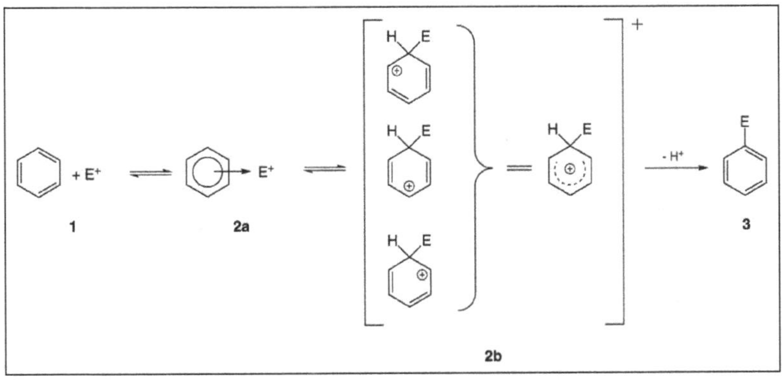


Electrophilic aromatic substitution.

Electrophilic reactions to other unsaturated compounds than arenes generally lead to electrophilic addition rather than substitution.

## Radical Substitution

A radical substitution reaction involves radicals. An example is the Hunsdiecker reaction.

## Organometallic Substitution

Coupling reactions are a class of metal-catalyzed reactions involving an organometallic compound RM and an organic halide R′X that together react to form a compound of the type R-R′ with formation of a new carbon–carbon bond. Examples include the Heck reaction, Ullmann reaction, and Wurtz–Fittig reaction. Many variations exist.

## Substituted Compounds

Substituted compounds are chemical compounds where one or more hydrogen atoms of a core structure have been replaced with a functional group like alkyl, hydroxy, or halogen, or with larger substituent groups.

For example, benzene is a simple aromatic ring. Benzenes that have undergone substitution are a heterogeneous group of chemicals with a wide spectrum of uses and properties:

| Examples of substituted benzene compounds | | |
|---|---|---|
| compound | general formula | general structure |
| Benzene | $C_6H_6$ | 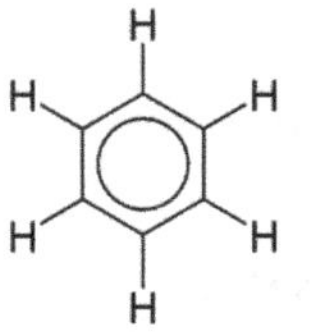 |
| Toluene | $C_6H_5\text{-}CH_3$ | 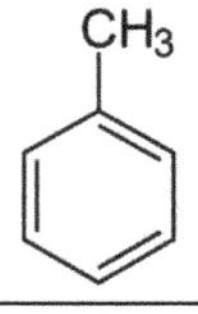 |
| o-Xylene | $C_6H_4(\text{-}CH_3)_2$ | 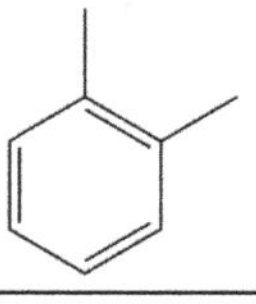 |
| Mesitylene | $C_6H_3(\text{-}CH_3)_3$ | 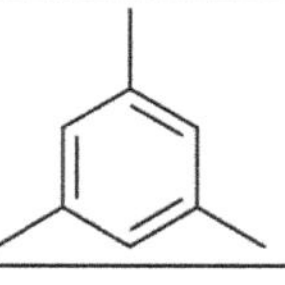 |
| Phenol | $C_6H_5\text{-}OH$ |  |

# Addition Reaction

An addition reaction, in organic chemistry, is in its simplest terms an organic reaction where two or more molecules combine to form a larger one (the adduct).

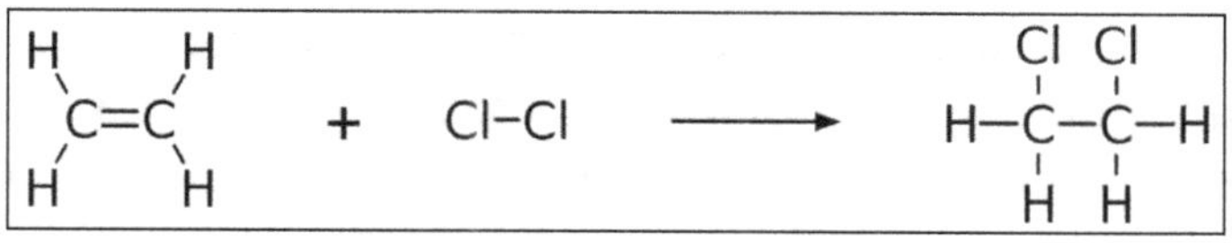

Addition of chlorine to ethylene.

Addition reactions are limited to chemical compounds that have multiple bonds, such as molecules with carbon–carbon double bonds (alkenes), or with triple bonds (alkynes). Molecules containing carbon—hetero double bonds like carbonyl (C=O) groups, or imine (C=N) groups, can undergo addition, as they too have double-bond character.

An addition reaction is the reverse of an elimination reaction. For instance, the hydration of an alkene to an alcohol is reversed by dehydration.

There are two main types of polar addition reactions: electrophilic addition and nucleophilic addition. Two non-polar addition reactions exist as well, called free-radical addition and cycloadditions. Addition reactions are also encountered in polymerizations and called addition polymerization.

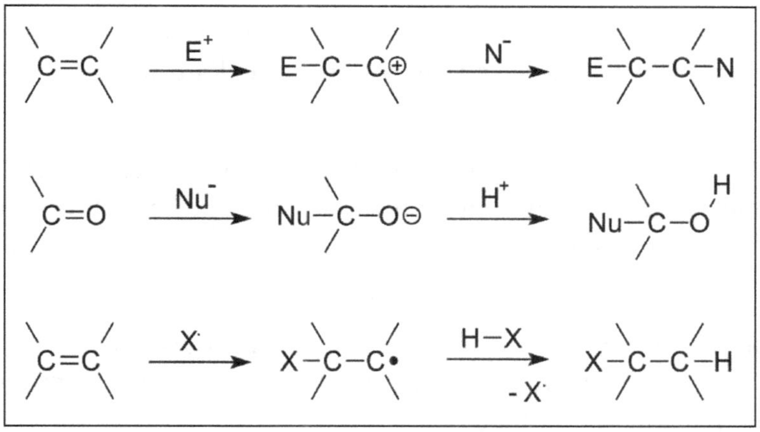

General overview of addition reactions. Top to bottom: electrophilic addition to alkene, nucleophilic addition of nucleophile to carbonyl and free-radical addition of halide to alkene.

Depending on the product structure, it could promptly react further to eject a leaving group to give the addition–elimination reaction sequence.

## Electrophilic Addition

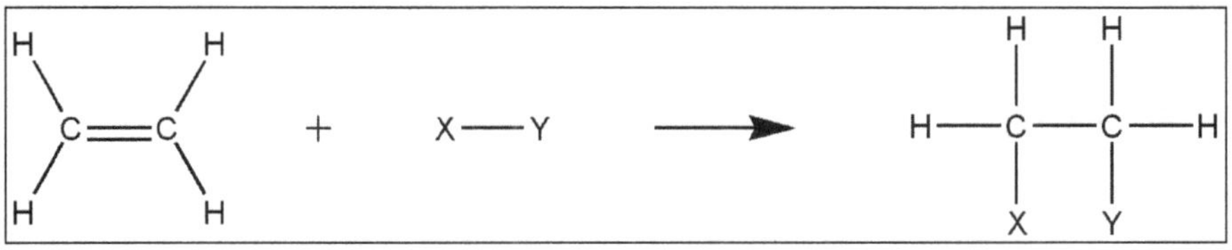

The overall reaction for electrophilic addition to ethylene.

In organic chemistry, an electrophilic addition reaction is an addition reaction where, in a chemical compound, a π bond is broken and two new σ bonds are formed. The substrate of an electrophilic addition reaction must have a double bond or triple bond.

The driving force for this reaction is the formation of an electrophile $X^+$ that forms a covalent bond with an electron-rich unsaturated C=C bond. The positive charge on X is transferred to the carbon-carbon bond, forming a carbocation during the formation of the C-X bond.

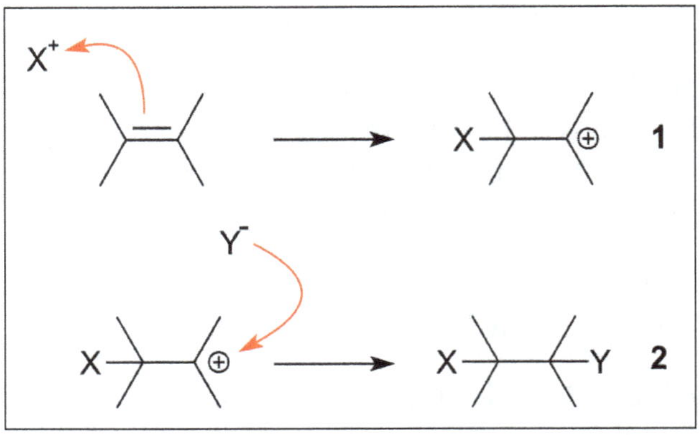


In step 2 of an electrophilic addition, the positively charged intermediate combines with (Y) that is electron-rich and usually an anion to form the second covalent bond.

Step 2 is the same nucleophilic attack process found in an $S_N1$ reaction. The exact nature of the electrophile and the nature of the positively charged intermediate are not always clear and depend on reactants and reaction conditions.

In all asymmetric addition reactions to carbon, regioselectivity is important and often determined by Markovnikov's rule. Organoborane compounds give anti-Markovnikov additions. Electrophilic attack to an aromatic system results in electrophilic aromatic substitution rather than an addition reaction.

In Electrophilic addition the electrophile with the positive charge effects the formation of the total structure, which thus bears a positive charge as well, to make up for the new addition, which then results in the intermediate, bearing that positive charge.

## Typical Electrophilic Additions

Typical electrophilic additions to alkenes with reagents are:

- Dihalo addition reactions: $X_2$
- Hydrohalogenations: HX
- Hydration reactions: $H_2O$
- Hydrogenations: $H_2$
- Oxymercuration reactions: mercuric acetate, water
- Hydroboration-oxidation reactions: diborane
- The Prins reaction: formaldehyde, water.

## Nucleophilic Addition

In organic chemistry, a nucleophilic addition reaction is an addition reaction where a chemical compound with an electron-deficient or electrophilic double or triple bond, a π bond, reacts with electron-rich reactant, termed a nucleophile, with disappearance of the double bond and creation of two new single, or σ, bonds. The reactions are involved in the biological synthesis of compounds in the metabolism of every living organism, and are used by chemists in academia and industries such as pharmaceuticals to prepare most new complex organic chemicals, and so are central to organic chemistry. Addition reactions require the presence of groups with multiple bonds in the electrophile(due to the fact that double bonds and even triple bonds can both lack electron rich sources): carbon–heteroatom multiple bonds as in carbonyls, imines, and nitriles, or carbon–carbon double or triple bonds. The lack of electron rich sources is due to the fact that these bonds are partially empty, even though they remain connected, since the region occupying the orbital is essentially dead. This electrophilic behavior is defined as empty space since everything inside is basically without any source of electricity except from outside the bond, since bonds tend to want to attract more to themselves(whether this be electric or non-electric can differ in most situations). The addition of the nucleophile means the continuous addition of a negative charge throughout the reaction, unless an electrophile also makes itself present to form a complete structure with no charge at all. The negative charge being continuous throughout the reaction until the formation of an intermediate, bearing the charge, thus is the addition reaction we have before us. Once this meets an electrophile, then the intermediate formed with the negative charge can thus be neutralized to form a complete structure via a type of bond.

### Addition to Carbon–heteroatom Double Bonds

Nucleophilic addition reactions of nucleophiles with electrophilic double or triple bond (π bonds) create a new carbon center with two additional single, or σ, bonds. Addition of a nucleophile to carbon–heteroatom double or triple bonds such as >C=O or -C≡N show great variety. These types of bonds are polar (have a large difference in electronegativity between the two atoms); consequently, their carbon atoms carries a partial positive charge. This makes the molecule an electrophile, and the carbon atom the electrophilic center; this atom is the primary target for the nucleophile. Chemists have developed a geometric system to describe the approach of the nucleophile to the electrophilic center, using two angles, the Bürgi–Dunitz and the Flippin–Lodge angles after scientists that first studied and described them.

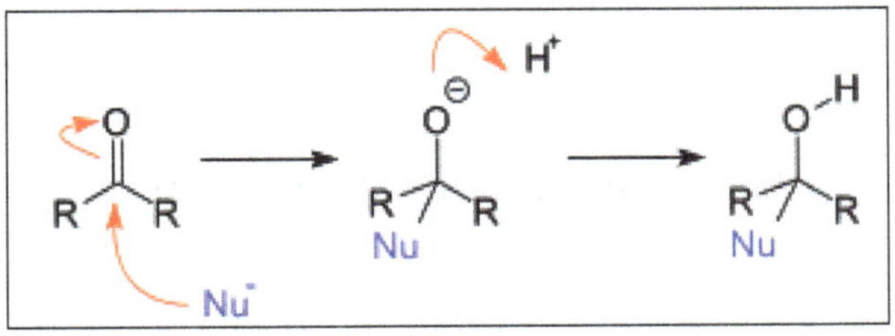

This type of reaction is also called a 1,2 nucleophilic addition. The stereochemistry of this type of nucleophilic attack is not an issue, when both alkyl substituents are dissimilar and there are not any other controlling issues such as chelation with a Lewis acid, the reaction product is a racemate. Addition reactions of this type are numerous. When the addition reaction is accompanied by an elimination the reaction type is nucleophilic acyl substitution or an addition-elimination reaction.

## Addition to Carbonyl Groups

With a carbonyl compound as an electrophile, the nucleophile can be:

- Water in hydration to a geminal diol (hydrate).
- An alcohol in acetalisation to an acetal.
- A hydride in reduction to an alcohol.
- An amine with formaldehyde and a carbonyl compound in the mannich reaction.
- An enolate ion in an aldol reaction or baylis–hillman reaction.
- An organometallic nucleophile in the grignard reaction or the related barbier reaction or a reformatskii reaction.
- Ylides such as a wittig reagent or the corey–chaykovsky reagent or α-silyl carbanions in the peterson olefination.
- A phosphonate carbanion in the horner–wadsworth–emmons reaction.
- A pyridine zwitterion in the hammick reaction.
- An acetylide in alkynylation reactions.
- A cyanide ion in cyanohydrin reactions.

In many nucleophilic reactions, addition to the carbonyl group is very important. In some cases, the C=O double bond is reduced to a C-O single bond when the nucleophile bonds with carbon. For example, in the cyanohydrin reaction a cyanide ion forms a C-C bond by breaking the carbonyl's double bond to form a cyanohydrin.

## Addition to Nitriles

With nitrile electrophiles, nucleophilic addition take place by:

- Hydrolysis of a nitrile to form an amide or a carboxylic acid.
- Organozinc nucleophiles in the blaise reaction.
- Alcohols in the pinner reaction.
- The (same) nitrile α-carbon in the thorpe reaction. The intramolecular version is called the thorpe–ziegler reaction.
- Grignard reagents to form imines. The route affords ketones following hydrolysis or primary amines following imine reduction.

## Addition to Carbon–Carbon Double Bonds

The driving force for the addition to alkenes is the formation of a nucleophile $X^-$ that forms a covalent bond with an electron-poor unsaturated system -C=C- (step 1). The negative charge on X is transferred to the carbon – carbon bond.

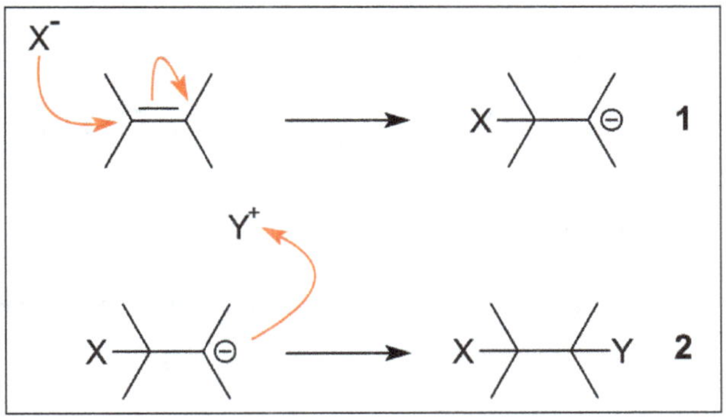


In step 2 the negatively charged carbanion combines with (Y) that is electron-poor to form the second covalent bond. Ordinary alkenes are not susceptible to a nucleophilic attack (apolar bond). Styrene reacts in toluene with sodium to 1,3-diphenylpropane through the intermediate carbanion:

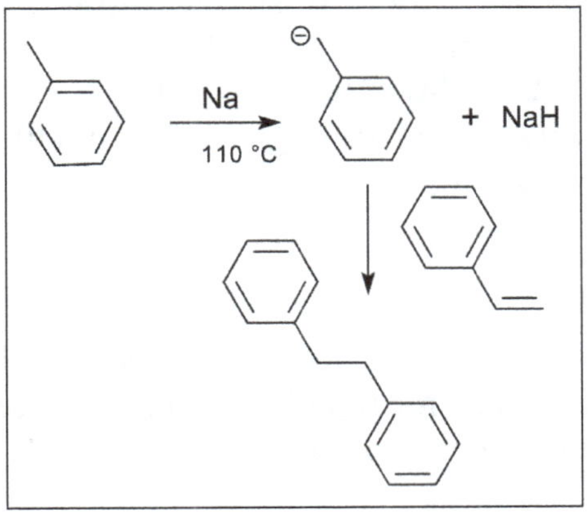


Another exception to the rule is found in the Varrentrapp reaction. Fullerenes have unusual double bond reactivity and additions such has the Bingel reaction are more frequent. When X is a carbonyl group like C=O or COOR or a cyanide group (CN), the reaction type is a conjugate addition reaction. The substituent X helps to stabilize the negative charge on the carbon atom by its inductive effect. In addition when Y-Z is an active hydrogen compound the reaction is known as a Michael reaction. Perfluorinated alkenes (alkenes that have all hydrogens replaced by fluorine) are highly prone to nucleophilic addition, for example by fluoride ion from caesium fluoride or silver(I) fluoride to give a perfluoroalkyl anion.

## Elimination Reaction

An elimination reaction is a type of organic reaction in which two substituents are removed from a molecule in either a one or two-step mechanism. The one-step mechanism is known as the E2 reaction, and the two-step mechanism is known as the E1 reaction. The numbers refer not to the number of steps in the mechanism, but rather to the kinetics of the reaction: E2 is bimolecular (second-order) while E1 is unimolecular (first-order). In cases where the molecule is able to stabilize an anion but possesses a poor leaving group, a third type of reaction, $E1_{CB}$, exists. Finally, the pyrolysis of xanthate and acetate esters proceed through an "internal" elimination mechanism, the $E_i$ mechanism.

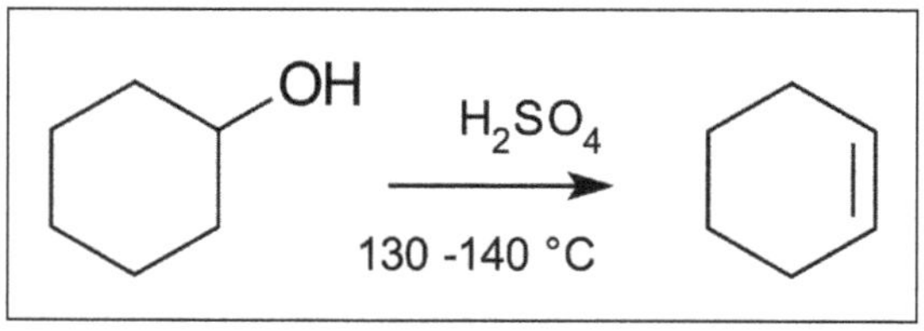


Elimination reaction of cyclohexanol to cyclohexene with sulfuric acid and heat.

## Loss of Hydron ($H^+$)

In most organic elimination reactions, at least one hydron ($H^+$) is lost to form the double bond: the unsaturation of the molecule increases. It is also possible that a molecule undergoes reductive elimination, by which the valence of an atom in the molecule decreases by two, though this is more common in inorganic chemistry. An important class of elimination reactions is those involving alkyl halides, with good leaving groups, reacting with a Lewis base to form an alkene. Elimination may be considered the reverse of an addition reaction. When the substrate is asymmetric, regioselectivity is determined by Zaitsev's rule or through Hofmann elimination if the carbon with the most substituted hydrogen is inaccessible.

## E2 Mechanism

During the 1920s, Sir Christopher Ingold proposed a model to explain a peculiar type of chemical reaction: the E2 mechanism. E2 stands for bimolecular elimination. The reaction involves a one-step mechanism in which *carbon-hydrogen* and *carbon-halogen* bonds break to form a double bond (*C=C Pi bond*).

The specifics of the reaction are as follows:

- E2 is a single step elimination, with a single transition state.
- It is typically undergone by primary substituted alkyl halides, but is possible with some secondary alkyl halides and other compounds.
- The reaction rate is second order, because it's influenced by both the alkyl halide and the base (bimolecular).
- Because the E2 mechanism results in the formation of a pi bond, the two leaving groups (often a hydrogen and a halogen) need to be antiperiplanar. An antiperiplanar transition state has staggered conformation with lower energy than a synperiplanar transition state which is in eclipsed conformation with higher energy. The reaction mechanism involving staggered conformation is more favorable for E2 reactions (unlike E1 reactions).
- E2 typically uses a strong base. It must be strong enough to remove a weakly acidic hydrogen.
- In order for the pi bond to be created, the hybridization of carbons needs to be lowered from $sp^3$ to $sp^2$.

- The C-H bond is weakened in the rate determining step and therefore a primary deuterium isotope effect much larger than 1 (commonly 2-6) is observed.
- E2 competes with the $S_N2$ reaction mechanism if the base can also act as a nucleophile (true for many common bases).

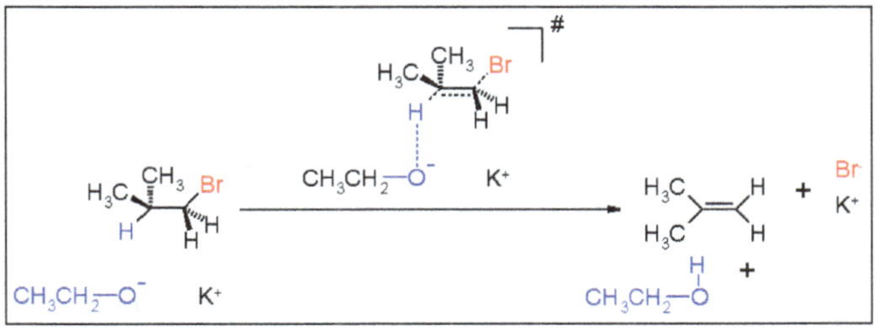

An example of this type of reaction in *scheme 1* is the reaction of isobutylbromide with potassium ethoxide in ethanol. The reaction products are isobutylene, ethanol and potassium bromide.

## E1 Mechanism

E1 is a model to explain a particular type of chemical elimination reaction. E1 stands for unimolecular elimination and has the following specifications:

- It is a two-step process of elimination: *ionization and deprotonation.*
    - Ionization: the carbon-halogen bond breaks to give a carbocation intermediate.
    - Deprotonation of the carbocation.
- E1 typically takes place with tertiary alkyl halides, but is possible with some secondary alkyl halides.
- The reaction rate is influenced only by the concentration of the alkyl halide because carbocation formation is the slowest step, as known as the rate-determining step. Therefore, first-order kinetics apply (unimolecular).
- The reaction usually occurs in the complete absence of a base or the presence of only a weak base (acidic conditions and high temperature).
- E1 reactions are in competition with $S_N1$ reactions because they share a common carbocationic intermediate.
- A secondary deuterium isotope effect of slightly larger than 1 (commonly 1 - 1.5) is observed.
- There is no antiperiplanar requirement. An example is the pyrolysis of a certain sulfonate ester of menthol:

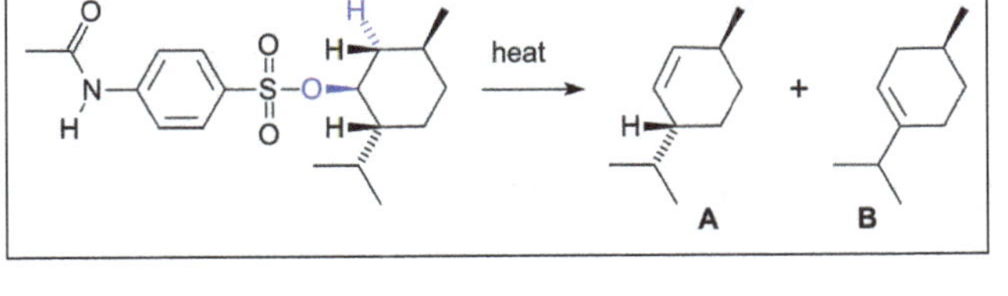

Only reaction product A results from antiperiplanar elimination. The presence of product B is an indication that an E1 mechanism is occurring.

- It is accompanied by carbocationic rearrangement reactions.

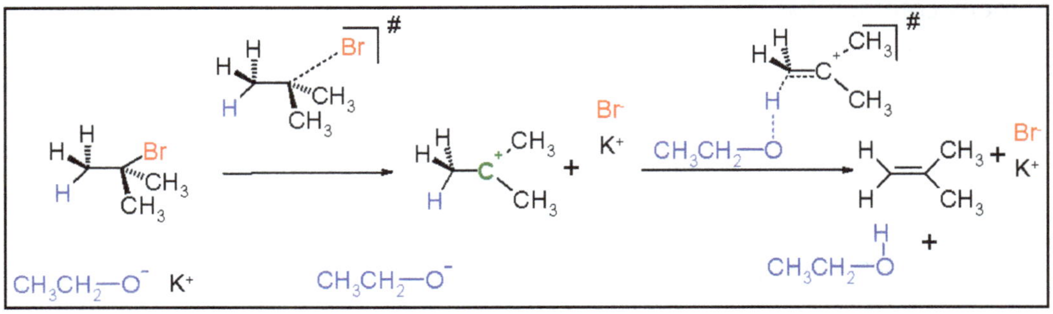

An example in scheme above is the reaction of tert-butylbromide with potassium ethoxide in ethanol.

E1 eliminations happen with highly substituted alkyl halides for two main reasons:

- Highly substituted alkyl halides are bulky, limiting the room for the E2 one-step mechanism; therefore, the two-step E1 mechanism is favored.
- Highly substituted carbocations are more stable than methyl or primary substituted cations. Such stability gives time for the two-step E1 mechanism to occur.
- If $S_N1$ and E1 pathways are competing, the E1 pathway can be favored by increasing the heat.

Specific features:

1. Rearrangement possible
2. Independent of concentration
3. Basicity of base

## Competition among Mechanisms

The reaction rate is influenced by the reactivity of halogens, iodide and bromide being favored. Fluoride is not a good leaving group, so eliminations with fluoride as the leaving group have slower rates than other halogens. There is a certain level of competition between the elimination reaction and nucleophilic substitution. More precisely, there are competitions between E2 and $S_N2$ and also between E1 and $S_N1$. Substitution generally predominates and elimination occurs only during precise circumstances. Generally, elimination is favored over substitution when:

- Steric hindrance around the α-carbon increases.
- A stronger base is used.
- Temperature increases (increase entropy)

- The base is a poor nucleophile. Bases with steric bulk, (such as in potassium tert-butoxide), are often poor nucleophiles.

In one study the kinetic isotope effect (KIE) was determined for the gas phase reaction of several alkyl halides with the chlorate ion. In accordance with an E2 elimination the reaction with t-butyl chloride results in a KIE of 2.3. The methyl chloride reaction (only $S_N2$ possible) on the other hand has a KIE of 0.85 consistent with a $S_N2$ reaction because in this reaction type the C-H bonds tighten in the transition state. The KIE's for the ethyl (0.99) and isopropyl (1.72) analogues suggest competition between the two reaction modes.

## Elimination Reactions other than β-elimination

β-Elimination, with loss of electrofuge and nucleofuge on vicinal carbons, is by far the most common type of elimination. The ability to form a stable product containing a C=C or C=X bond, as well as orbital alignment considerations, strongly favors β-elimination over other elimination processes. However, other types are known, generally for systems where β-elimination cannot occur.

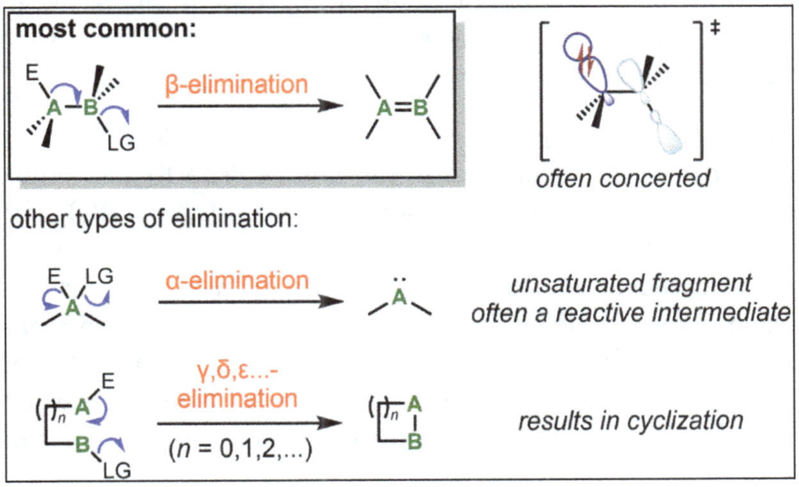


The next most common type of elimination reaction is α-elimination. For a carbon center, the result of α-elimination is the formation of a carbene, which includes "stable carbenes" such as carbon monoxide or isocyanides. For instance, α-elimination the elements of HCl from chloroform ($CHCl_3$) in the presence of strong base is a classic approach for the generation of dichlorocarbene, $:CCl_2$, as a reactive intermediate. On the other hand, formic acid undergoes α-elimination to afford the stable products water and carbon monoxide under acidic conditions. α-Elimination may also occur on a metal center, one particularly common result of which is lowering of both the metal oxidation state and coordination number by 2 units in a process known as reductive elimination. (Confusingly, in organometallic terminology, the terms *α-elimination* and *α-abstraction* refer to processes that result in formation of a metal-carbene complex. In these reactions, it is the carbon adjacent to the metal that undergoes α-elimination).

In certain special cases, γ- and higher eliminations to form three-membered or larger rings is also possible in both organic and organometallic processes. For instance, certain Pt(II) complexes undergo γ- and δ-elimination to give metallocycles. More recently, γ-silyl elimination of a silylcyclobutyl tosylate has been used to prepare strained bicyclic systems.

## E1cB-elimination Reaction

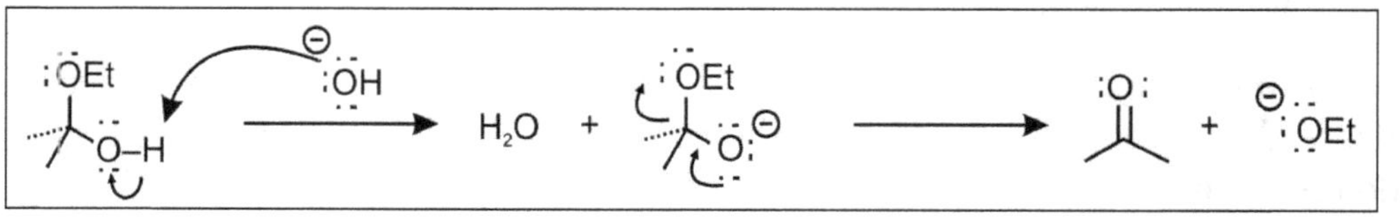


An example of the E1cB reaction mechanism in the degradation of a hemiacetal under basic conditions.

The E1cB elimination reaction is a type of elimination reaction which occurs under basic conditions, where the hydrogen to be removed is relatively acidic, while the leaving group (such as -OH or -OR) is a relatively poor one. Usually a moderate to strong base is present. E1cB is a two-step process, the first step of which may or may not be reversible. First, a base abstracts the relatively acidic proton to generate a stabilized anion. The lone pair of electrons on the anion then moves to the neighboring atom, thus expelling the leaving group and forming double or triple bond. The name of the mechanism - E1cB - stands for Elimination Unimolecular conjugate Base. *Elimination* refers to the fact that the mechanism is an elimination reaction and will lose two substituents. *Unimolecular* refers to the fact that the rate-determining step of this reaction only involves one molecular entity. Finally, *conjugate base* refers to the formation of the carbanion intermediate, which is the conjugate base of the starting material.

E1cb should be thought of as being on one end of a continuum, which includes the E1 mechanism at the opposite end, and the E2 mechanism in the middle. The E1 mechanism usually has the opposite characteristics: the leaving group is a good one (like -OTs or -Br), while the hydrogen is not particularly acidic and a strong base is absent. Thus, in the E1 mechanism, the leaving group leaves first to generate a carbocation. Due to the presence of an empty p orbital after departure of the leaving group, the hydrogen on the neighboring carbon becomes much more acidic, allowing it to then be removed by the weak base in the second step. In an E2 reaction, the presence of a strong base and a good leaving group allows proton abstraction by the base and the departure of the leaving group to occur simultaneously, leading to a concerted transition state in a one-step process.

## Mechanism

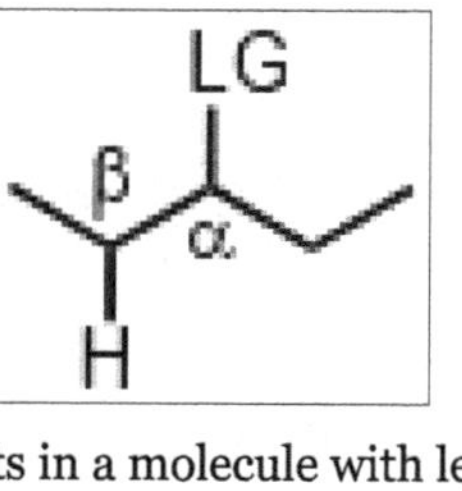


α and β assignments in a molecule with leaving group, LG.

There are two main requirements to have a reaction proceed down an E1cB mechanistic pathway. The compound must have an acidic hydrogen on its β-carbon and a relatively poor leaving group on the α- carbon. The first step of an E1cB mechanism is the deprotonation of the β-carbon, resulting in the formation of an anionic transition state, such as a carbanion. The greater the stability of this transition state, the more the mechanism will favor an E1cB mechanism. This transition state can be stabilized through induction or delocalization of the electron lone pair through resonance. In general it can be claimed that an electron withdrawing group on the substrate, a strong base, a poor leaving group and a polar solvent triggers the E1cB mechanism. An example of an E1cB mechanism that has a stable transition state can be seen in the degradation of ethiofencarb - a carbamate insecticide that has a relatively short half-life in earth's atmosphere. Upon deprotonation of the amine, the resulting amide is relatively stable because it is conjugated with the neighboring carbonyl. In addition to containing an acidic hydrogen on the β-carbon, a relatively poor leaving group is also necessary. A bad leaving group is necessary because a good leaving group will leave before the ionization of the molecule. As a result, the compound will likely proceed through an E2 pathway. Some examples of compounds that contain poor leaving groups and can undergo the E1cB mechanism are alcohols and fluoroalkanes. It has also been suggested that the E1cB mechanism is more common among alkenes eliminating to alkynes than from an alkane to alkene. One possible explanation for this is that the $sp^2$ hybridization creates slightly more acidic protons. Although this mechanism is not limited to carbon-based eliminations. It has been observed with other heteroatoms, such as nitrogen in the elimination of a phenol derivative from ethiofencarb.

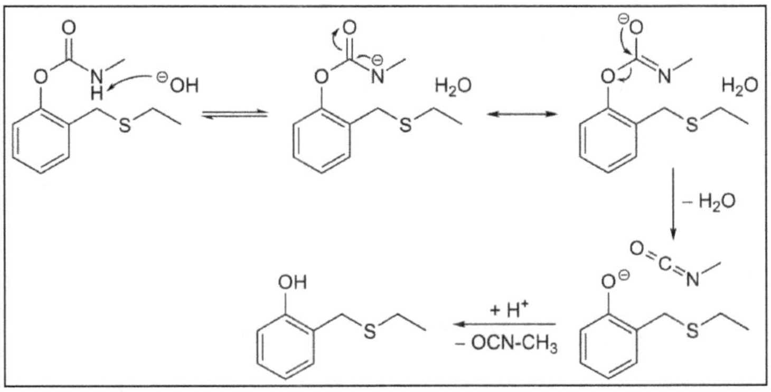


Degradation of ethiofencarb illustrating the presence of a stable anion due to resonance between the amide functional group and the carbonyl group.

## E1cB-elimination Reactions vs. E1- and E2-elimination Reactions

All elimination reactions involve the removal of two substituents from a pair of adjacent atoms in a compound. Alkene, alkynes, or similar heteroatom variations (such as carbonyl and cyano) will form. The E1cB mechanism is just one of three types of elimination reaction. The other two elimination reactions are E1 and E2 reactions. Although the mechanisms are similar, they vary in the timing of the deprotonation of the α-carbon and the loss of the leaving group. E1 stands for unimolecular elimination, and E2 stands for bimolecular elimination. In an E1 mechanism, the molecule contains a good leaving group that departs before deprotonation of the α-carbon. This results in the formation of a carbocation intermediate. The carbocation is then deprotonated resulting in the formation of a new pi bond. The molecule involved must also have a very good leaving group such as bromine or chlorine, and it should have a relatively less acidic α-carbon.

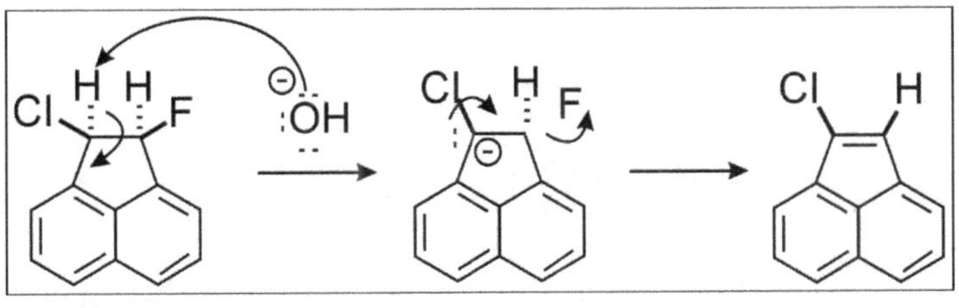


Example of the preferential elimination of fluorine in an E1cB-elimination reaction.

In an E2-elimination reaction, both the deprotonation of the α-carbon and the loss of the leaving group occur simultaneously in one concerted step. Molecules that undergo E2-elimination mechanisms have more acidic α-carbons than those that undergo E1 mechanisms, but their α-carbons are not as acidic as those of molecules that undergo E1cB mechanisms. The key difference between the E2 vs E1cb pathways is a distinct carbanion intermediate as opposed to one concerted mechanism. Studies have been shown that the pathways differ by using different halogen leaving groups. One example uses chlorine as a better stabilizing halogen for the anion than fluorine, which makes fluorine the leaving group even though chlorine is a much better leaving group. This provides evidence that the carbanion is formed because the products are not possible through the most stable concerted E2 mechanism. The following table summarizes the key differences between the three elimination reactions; however, the best way to identify which mechanism is playing a key role in a particular reaction involves the application of chemical kinetics.

| E1 | E2 | E1cB |
|---|---|---|
| Stepwise reaction | Concerted reaction | Stepwise reaction |
| Carbocation intermediate | Simultaneous removal of proton, formation of double bond, and loss of leaving group | Carbanion intermediate |
| No kind of conclusion | No preference | No kind of conclusion |
| Good leaving groups | Leaving group | Poor leaving groups |
| Less acidic B-H | Acidic B-H | More acidic B-H |

## Chemical Kinetics of E1cB-elimination Mechanisms

When trying to determine whether or not a reaction follows the E1cB mechanism, chemical kinetics are essential. The best way to identify the E1cB mechanism involves the use of rate laws and the kinetic isotope effect. These techniques can also help further differentiate between E1cB, E1, and E2-elimination reactions.

### Rate Law

When trying to experimentally determine whether or not a reaction follows the E1cB mechanism, chemical kinetics are essential. The best ways to identify the E1cB mechanism involves the use of rate laws and the kinetic isotope effect.

The rate law that governs E1cB mechanisms is relatively simple to determine. Consider the following reaction scheme.

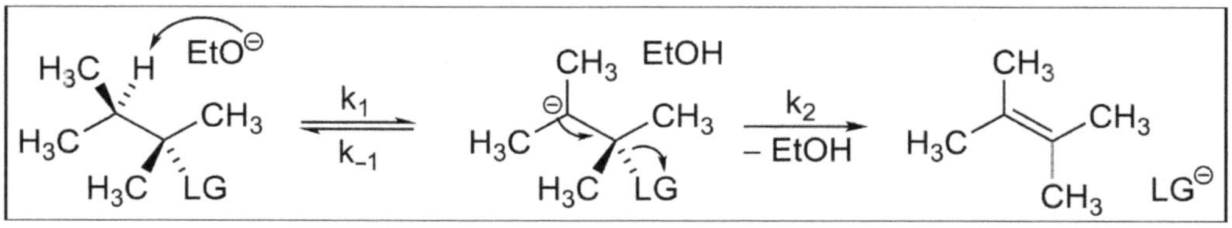


An example of an E1cB-elimination mechanism with a generic leaving group, LG, and ethoxide as the base.

Assuming that there is a steady-state carbanion concentration in the mechanism, the rate law for an E1cB mechanism.

$$\frac{d[P]}{dt} = \frac{k_1 k_2 [\text{substrate}][\text{base}]}{k_{-1}[\text{conjugate acid}] + k_2}$$

$$\text{rate} = k_{obs}[\text{substrate}][\text{base}]$$

From this equation, it is clear the second order kinetics will be exhibited. E1cB mechanisms kinetics can vary slightly based on the rate of each step. As a result, the E1cB mechanism can be broken down into three categories:

1. $E1cB_{anion}$ is when the carbanion is stable and/or a strong base is used in excess of the substrate, making deprotonation irreversible, followed by rate-determining loss of the leaving group ($k_1[\text{base}] \gg k_2$).
2. $E1cB_{rev}$ is when the first step is reversible but the formation of product is slower than reforming the starting material, this again results from a slow second step ($k_{-1}[\text{conjugate acid}] \gg k_2$).
3. $E1cB_{irr}$ is when the first step is slow, but once the anion is formed the product quickly follows ($k_2 \gg k_{-1}[\text{conjugate acid}]$). This leads to an irreversible first step but unlike $E1cB_{anion}$, deprotonation is rate determining.

## Deuterium

Deuterium exchange and a deuterium kinetic isotope effect can help distinguish among $E1cB_{rev}$, $E1cB_{anion}$, and $E1cB_{irr}$. If the solvent is protic and contains deuterium in place of hydrogen (e.g., $CH_3OD$), then the exchange of protons into the starting material can be monitored. If the recovered starting material contains deuterium, then the reaction is most likely undergoing an $E1cB_{rev}$ type mechanism. Recall, in this mechanism protonation of the carbanion (either by the conjugate acid or by solvent) is faster than loss of the leaving group. This means after the carbanion is formed, it will quickly remove a proton from the solvent to form the starting material. If the reactant contains deuterium at the β position, a primary kinetic isotope effect indicates that deprotonation is rate determining. Of the three E1cB mechanisms, this result is only consistent with the $E1cB_{irr}$ mechanism, since the isotope is already removed in $E1cB_{anion}$ and leaving group departure is rate determining in $E1cB_{rev}$.

## Fluorine-19 and Carbon-11

Another way that the kinetic isotope effect can help distinguish E1cB mechanisms involves the use of $^{19}F$. Fluorine is a relatively poor leaving group, and it is often employed in E1cB mechanisms. Fluorine kinetic isotope effects are also applied in the labeling of Radiopharmaceuticals and other compounds in medical research. This experiment is very useful in determining whether or not the loss of the leaving group is the rate-determining step in the mechanism and can help distinguish between $E1cB_{irr}$ and E2 mechanisms. $^{11}C$ can also be used to probe the nature of the transition state structure. The use of $^{11}C$ can be used to study the formation of the carbanion as well as study its lifetime which can not only show that the reaction is a two-step E1cB mechanism (as opposed to the concerted E2 mechanism), but it can also address the lifetime and stability of the transition state structure which can further distinguish between the three different types of E1cB mechanisms.

## Aldol Reactions

The most well known reaction that undergoes E1cB elimination is the aldol condensation reaction under basic conditions. This involves the deprotonation of a compound containing a carbonyl group that results in the formation of an enolate. The enolate is the very stable conjugate base of the starting material, and is one of the intermediates in the reaction. This enolate then acts as a nucleophile and can attack an electrophilic aldehyde. The Aldol product is then deprotonated forming another enolate followed by the elimination of water in an E1cB dehydration reaction. Aldol reactions are a key reaction in organic chemistry because they provide a means of forming carbon-carbon bonds, allowing for the synthesis of more complex molecules.

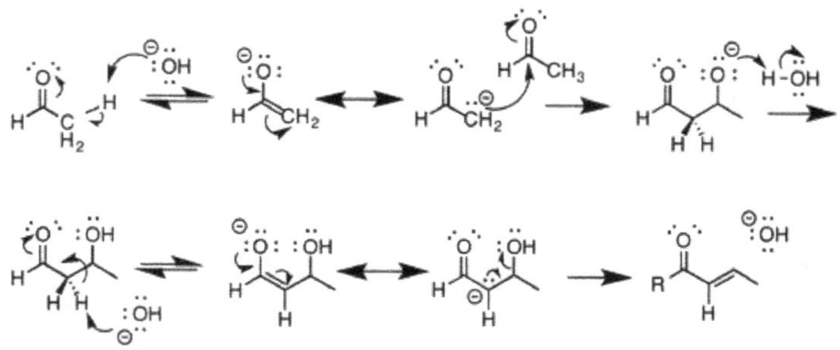

An aldol condensation reaction is one of the most common examples of an E1cB mechanism.

## Photo-induced E1cB

A photochemical version of E1cB has been reported by Lukeman *et al.* In this report, a photochemically induced decarboxylation reaction generates a carbanion intermediate, which subsequently eliminates the leaving group. The reaction is unique from other forms of E1cB since it does not require a base to generate the carbanion. The carbanion formation step is irreversible, and should thus be classified as $E1cB_{irr}$.

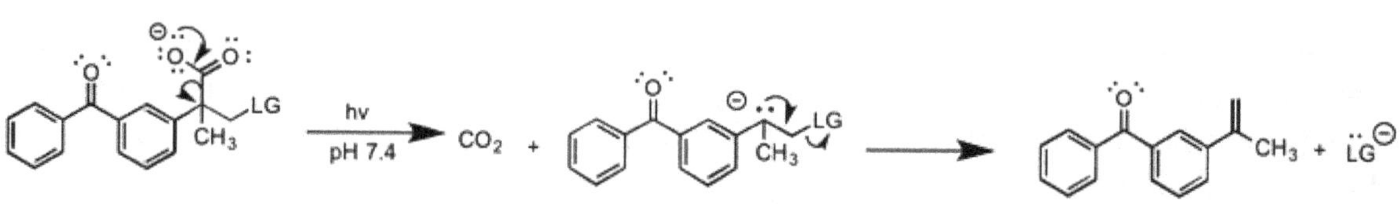

E1cB reaction mechanism through photo-induced decarboxylation.

# Rearrangement Reaction

A rearrangement reaction is a broad class of organic reactions where the carbon skeleton of a molecule is rearranged to give a structural isomer of the original molecule. Often a substituent moves from one atom to another atom in the same molecule. In the example below the substituent R moves from carbon atom 1 to carbon atom 2:

```
C–C–C– ⟶ –C–C–C
|              |
R              R
```

Intermolecular rearrangements also take place.

A rearrangement is not well represented by simple and discrete electron transfers (represented by curly arrows in organic chemistry texts). The actual mechanism of alkyl groups moving, as in Wagner-Meerwein rearrangement, probably involves transfer of the moving alkyl group fluidly along a bond, not ionic bond-breaking and forming. In pericyclic reactions, explanation by orbital interactions give a better picture than simple discrete electron transfers. It is, nevertheless, possible to draw the curved arrows for a sequence of discrete electron transfers that give the same result as a rearrangement reaction, although these are not necessarily realistic. In allylic rearrangement, the reaction is indeed ionic.

Three key rearrangement reactions are 1,2-rearrangements, pericyclic reactions and olefin metathesis.

## 1,2-rearrangements

A 1,2-rearrangement is an organic reaction where a substituent moves from one atom to another atom in a chemical compound. In a 1,2 shift the movement involves two adjacent atoms but moves over larger distances are possible. Skeletal isomerization is not normally encountered in the laboratory, but is the basis of large applications in oil refineries. In general straight-chain alkanes, are converted to branched isomers by heating in the presence of a catalyst. Examples include isomerisation of n-butane to isobutane and pentane to isopentane. Highly branched alkanes have favorable combustion characteristics for internal combustion engines.

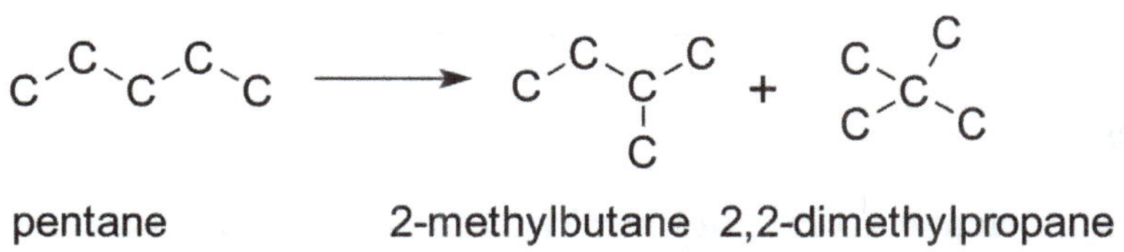


Further examples are the Wagner-Meerwein rearrangement:

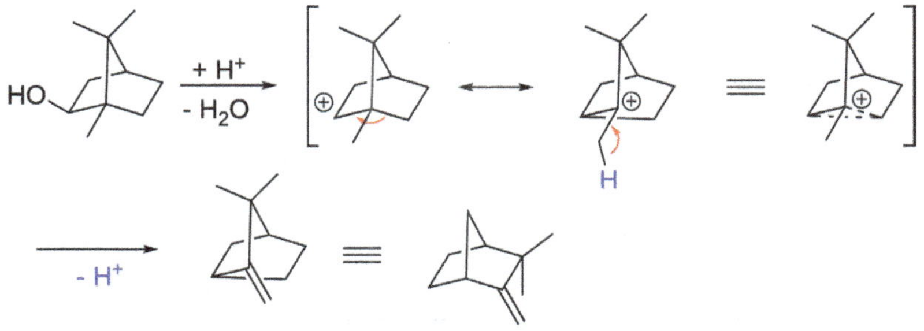


and the Beckmann rearrangement, which is relevant to the production of certain nylons:

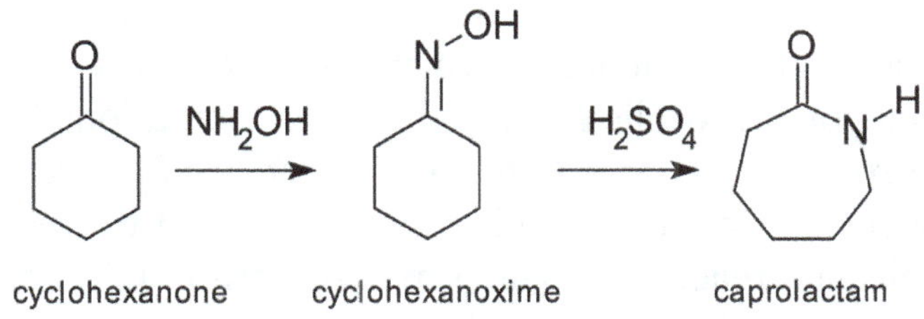

## Pericyclic Reactions

A pericyclic reaction is a type of reaction with multiple carbon-carbon bond making and breaking wherein the transition state of the molecule has a cyclic geometry, and the reaction progresses in a concerted fashion. Examples are hydride shifts and the Claisen rearrangement:

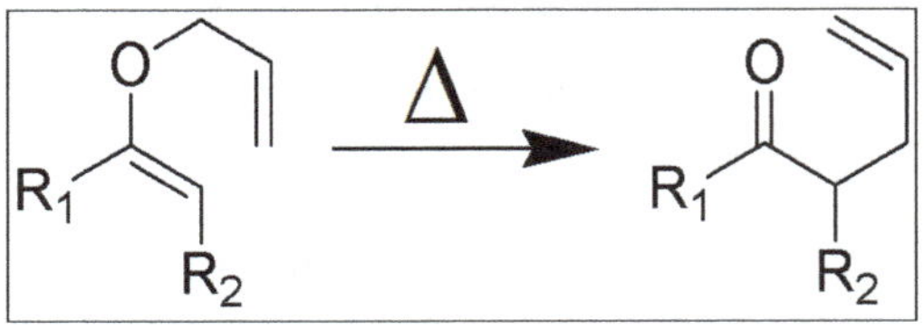

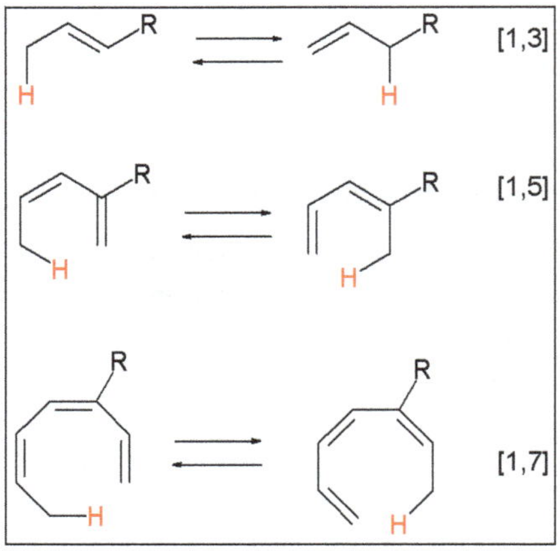

## Olefin Metathesis

Olefin metathesis is a formal exchange of the alkylidene fragments in two alkenes. It is a catalytic reaction with carbene, or more accurately, transition metal carbene complex intermediates.

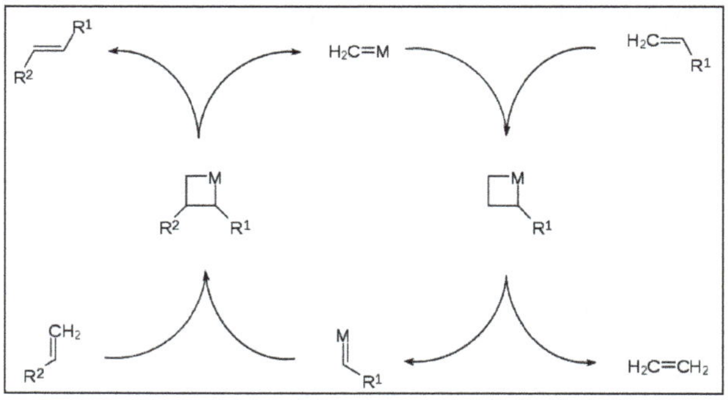

In this example (ethenolysis, a pair of vinyl compounds form a new symmetrical alkene with expulsion of ethylene.

## Schmidt Reaction

The Schmidt reaction is an organic reaction in which an azide reacts with a carbonyl derivative, usually a aldehyde, ketone, or carboxylic acid, under acidic conditions to give an amine or amide, with expulsion of nitrogen. It is named after Karl Friedrich Schmidt (1887–1971), who first reported it in 1924 by successfully converting benzophenone and hydrazoic acid to benzanilide. Surprisingly, the intramolecular reaction was not reported until 1991 but has become important in the synthesis of natural products.

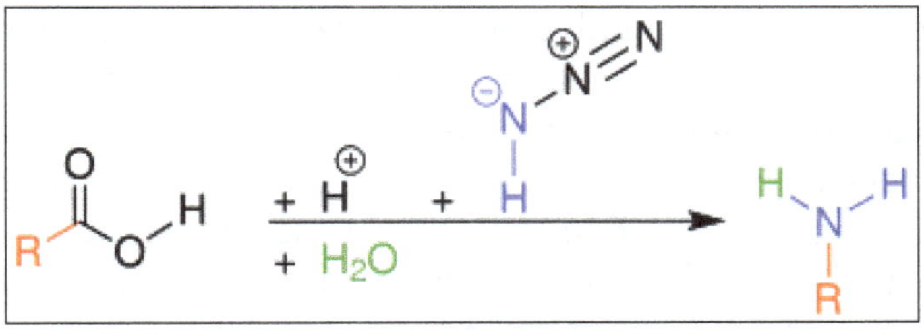

The reaction is effective with carboxylic acids to give amines (above), and with ketones to give amides (below).

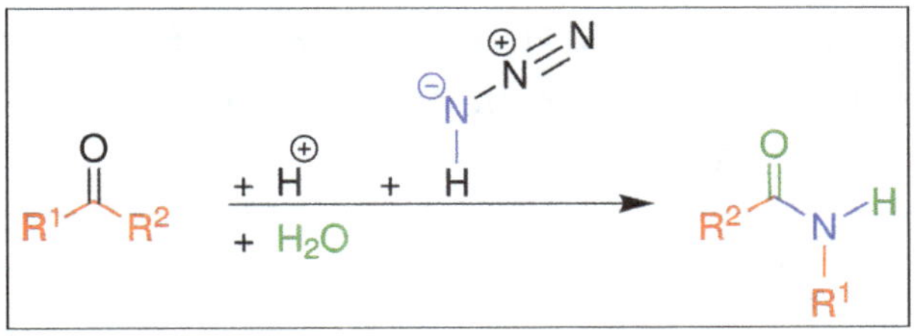

## Reaction Mechanism

The reaction is closely related to the Curtius rearrangement except that in this reaction the acyl azide is produced by reaction of the carboxylic acid with hydrazoic acid via the protonated carboxylic acid, in a process akin to a Fischer esterification. An alternative, involving the formation of an acylium ion, becomes more important when the reaction takes place in concentrated acid (>90% sulfuric acid). (In the Curtius rearrangement, sodium azide and an acyl chloride are combined to quantitatively generate the acyl azide intermediate, and the rest of the reaction takes place under neutral conditions.)

The carboxylic acid Schmidt reaction starts with acylium ion 1 obtained from protonation and loss of water. Reaction with hydrazoic acid forms the protonated azido ketone 2, which goes through a rearrangement reaction with the alkyl group R, migrating over the C-N bond with expulsion of nitrogen. The protonated isocyanate is attacked by water forming carbamate 4, which after deprotonation loses carbon dioxide to the amine.

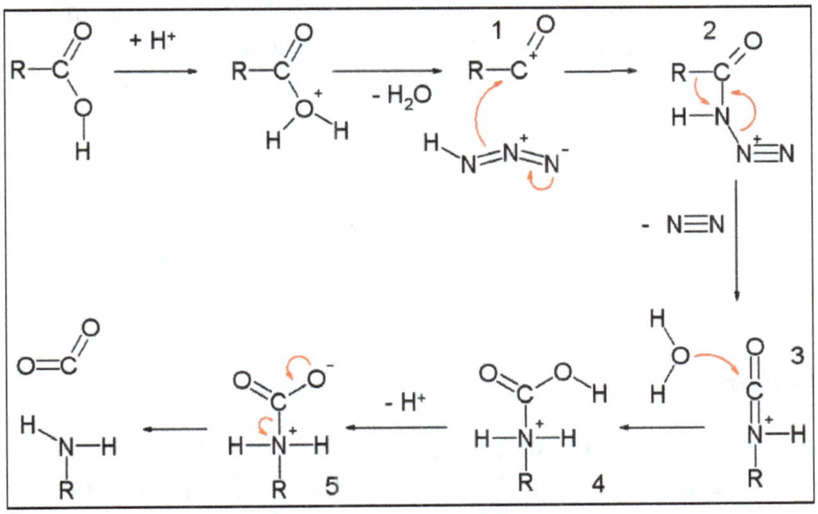

Schmidt reaction mechanism amine formation.

In the reaction mechanism for the Schmidt reaction of ketones, the carbonyl group is activated by protonation for nucleophilic addition by the azide, forming azidohydrin 3, which loses water in an elimination reaction to diazoiminium 5. One of the alkyl or aryl groups migrates from carbon to nitrogen with loss of nitrogen to give a nitrilium intermediate 6, as in the Beckmann rearrangement. Attack by water converts 6 to protonated imidic acid 7, which undergoes loss of proton to arrive at the imidic acid tautomer of the final amide. In an alternative mechanism, the migration occurs at 9, directly after protonation of intermediate 3, in a manner similar to the Baeyer-Villiger reaction to give protonated amide 10. Loss of a proton again furnishes the amide. It has been proposed that the dehydration to 3 to give 5 (and, hence, the Beckmann pathway) is favored by nonaqueous acids like conc. $H_2SO_4$, while aqueous acids like conc. HCl favor migration from 9 (the Baeyer-Villiger pathway). These possibilities have been used to account for the fact that, for certain substrates like α-tetralone, the group that migrates can sometimes change, depending on the conditions used, to deliver either of the two possible amides.

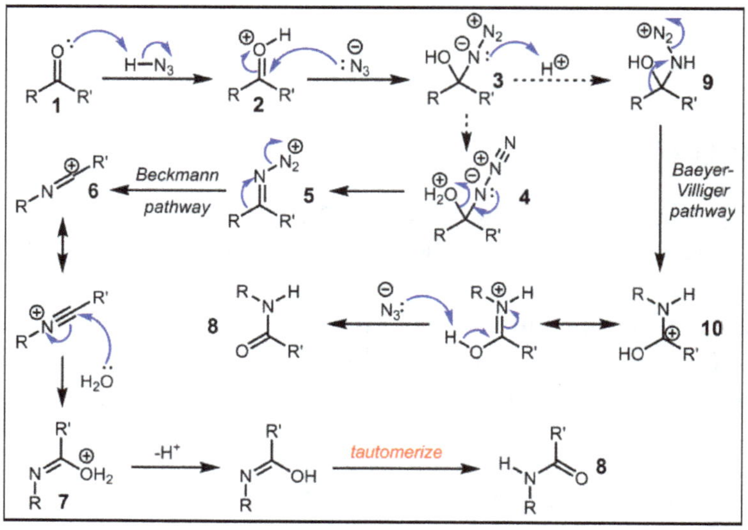


## Reactions Involving Alkyl Azides

The scope of this reaction has been extended to reactions of carbonyls with alkyl azides R-$N_3$. This extension was first reported by J.H. Boyer in 1955 (hence the name Boyer reaction), for exam-ple, the reaction of m-nitrobenzaldehyde with β-azido-ethanol:

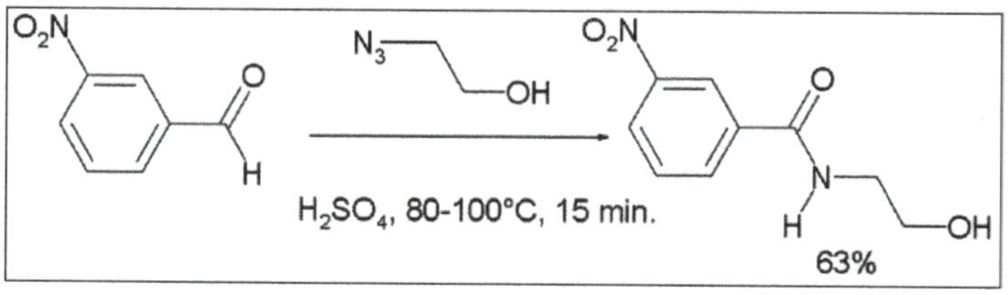


The Boyer reaction.

## The Boyer Reaction

Variations involving intramolecular Schmidt reactions have been known since 1991. These are annulation reactions and have some utility in the synthesis of natural products; such as lactams and alkaloids.

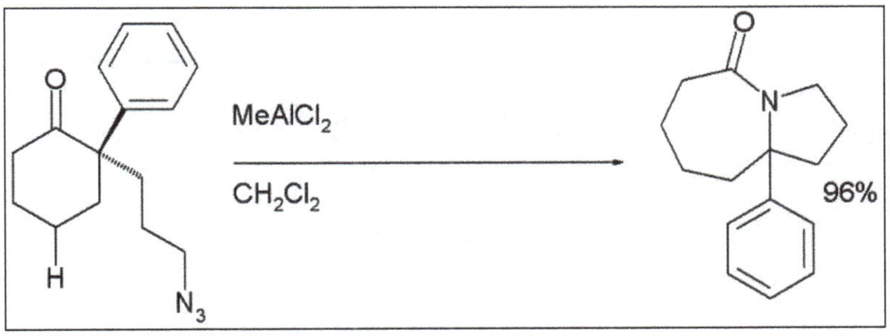


Intramolecular Schmidt Reaction.

## Wolff Rearrangement

The Wolff rearrangement is a reaction in organic chemistry in which an α-diazocarbonyl compound is converted into a ketene by loss of dinitrogen with accompanying 1,2-rearrangement. The Wolff rearrangement yields a ketene as an intermediate product, which can undergo nucleophilic attack with weakly acidic nucleophiles such as water, alcohols, and amines, to generate carboxylic acid derivatives or undergo [2+2] cycloaddition reactions to form four-membered rings. The mechanism of the Wolff rearrangement has been the subject of debate since its first use. No single mechanism sufficiently describes the reaction, and there are often competing concerted and carbene-mediated pathways; for simplicity, only the textbook, concerted mechanism is shown below. The reaction was discovered by Ludwig Wolff in 1902. The Wolff rearrangement has great synthetic utility due to the accessibility of α-diazocarbonyl compounds, variety of reactions from the ketene intermediate, and stereochemical retention of the migrating group. However, the Wolff rearrangement has limitations due to the highly reactive nature of α-diazocarbonyl compounds, which can undergo a variety of competing reactions.

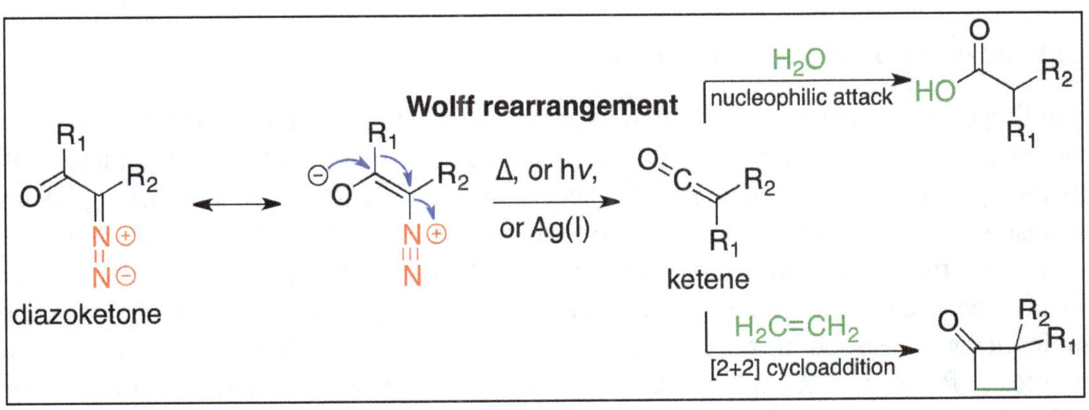


The Wolff rearrangement can be induced via thermolysis, photolysis, or transition metal catalysis. In this last case, the reaction is sensitive to the transition metal; silver (I) oxide or other Ag(I) catalysts work well and are generally used. The Wolff rearrangement has been used in many total syntheses; the most common use is trapping the ketene intermediate with nucleophiles to form carboxylic acid derivatives. The Arndt-Eistert homologation is a specific example of this use, wherein a carboxylic acid may be elongated by a methylene unit. Another common use is in ring-contraction methods; if the α-diazo ketone is cyclic, the Wolff rearrangement results in a ring-contracted product. The Wolff rearrangement works well in generating ring-strained systems, where other reactions may fail.

## Mechanism

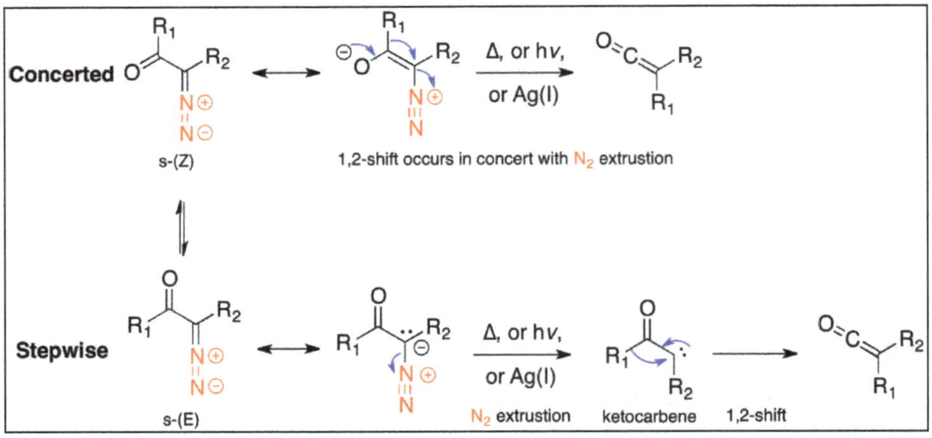


The mechanistic pathway of the Wolff-rearrangement has been the subject of much debate, as there are often competing concerted and stepwise mechanisms. However, two aspects of the mechanism can be agreed upon. First, α-diazocarbonyl compounds are in an equilibrium of s-*cis* and s-*trans*-conformers, the distribution of which may influence the mechanism of the reaction. Generally, under photolysis, compounds in the s-*cis* conformation react in a concerted manner due to the antiperiplanar relationship between the leaving and migrating groups, whereas compounds in the s-*trans* conformation react stepwise through a carbene intermediate or do not rearrange. Second, regardless of the reaction mechanism, the rearrangement gives a ketene intermediate, which can be trapped by a weakly acidic nucleophile, such as an alcohol or amine, to give the corresponding ester or amide, or an olefin, to give a [2+2] cycloaddition adduct. Strong acids do not rearrange, but rather protonate the α-carbon and give $S_N2$ products.

## Stereochemistry of α-diazo Ketones

Understanding the stereochemistry of α-diazo ketones is essential in elucidating the mechanism of the Wolff rearrangement. α-diazocarbonyl compounds are generally locally planar, with large rotational barriers (55–65 kJ/mol) due to C-C olefin character between the carbonyl and α-carbon, illustrated in the rightmost resonance structure. Such a large barrier slows molecular rotations sufficiently to lead to an equilibrium between two conformers, an s-*trans* and s-*cis*-conformer. s-*cis*-Conformers are electronically favored due to Coulombic attraction between the oxygen with a partial negative charge and the cationic nitrogen, as seen in the rightmost resonance structure. If $R_1$ is large and $R_2$ is hydrogen, s-*cis* is sterically favored. If $R_1$ and $R_2$ are large, s-*trans* is sterically favored; if both substitents are sufficiently large, the steric repulsion can outweigh the Coulombic attraction, leading to a preference for s-*trans*. Small and medium cyclic substrates are constrained in the s-*cis* conformation.

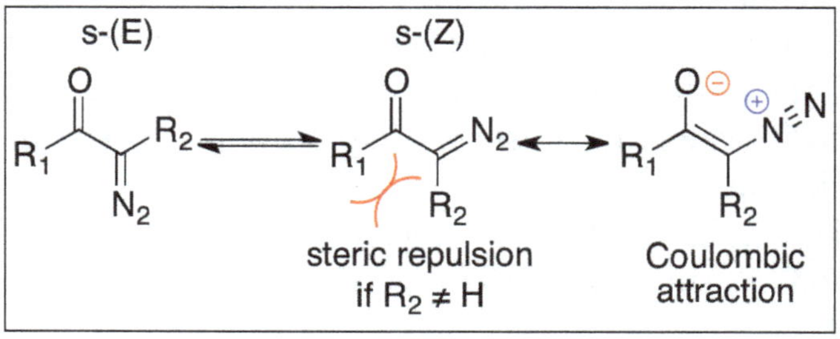

## Concerted Mechanism

When the α-diazo ketone is in the s-*cis* conformation, the leaving group ($N_2$) and the migrating group ($R_1$) are antiperiplanar, which favors a concerted mechanism, in which nitrogen extrusion occurs concurrently with 1,2-alkyl shift. There is evidence this mechanism occurs in both thermolytic and photolytic methods, when the s-*cis*-conformer is strongly favored.

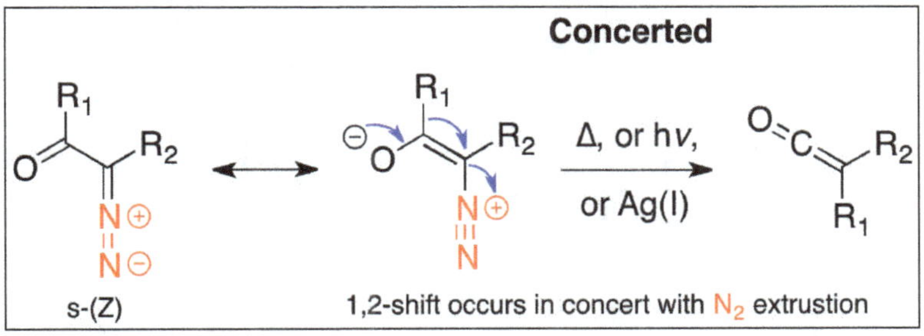


CIDNP studies show that photochemical rearrangement of diazoacetone, which largely exists in the s-*cis*-conformer, is concerted. Product ratios from direct and triplet-sensitized photolysis have been used as evidence for proposals that claim that concerted products arise from the s-*cis*-conformer and stepwise products occur through the s-*trans*-conformer.

## Stepwise Mechanism

s-*trans*-α-Diazo ketones do not have an antiperiplanar relationship between the leaving and migrating group, and thus are thought to generally rearrange stepwise. The stepwise mechanism begins with nitrogen extrusion, forming an α-ketocarbene. The α-ketocarbene can either undergo a 1,2-alkyl shift, to give the ketene product, or can undergo a 4π electrocyclic ring closure, to form an antiaromatic oxirene. This oxirene can reopen in two ways, to either α-ketocarbene, which can then form the ketene product.

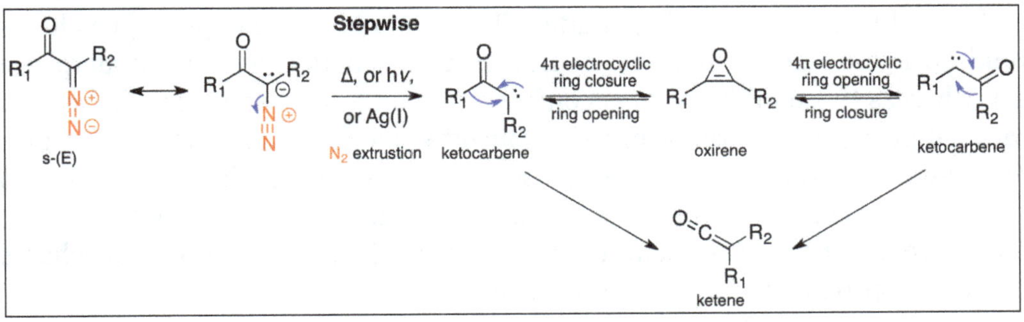


There are two primary arguments for stepwise mechanisms. The first is that rate constants of Wolff rearrangements depend on the stability of the formed carbene, rather than the migratory aptitude of the migrating group. The most definitive evidence is isotopic scrambling of the ketene, as predicted by an oxirene intermediate, which can only occur in the stepwise path. In the scheme below, the red carbon is $^{13}C$ labelled. The symmetric oxirene intermediate can open either way, scrambling the $^{13}C$ label. If the substituents $R_1$ and $R_2$ are the same, one can quantify the ratio of products stemming from the concerted and stepwise mechanisms; if the substituents are different, the oxirene will have a preference in the direction it opens, and a ratio cannot be quantified, but any scrambling indicates some reactant is going through a stepwise mechanism. In photolysis of diazo acetaldehyde, 8% of the label is scrambled, indicating that 16% of product is formed via the oxirene intermediate. Under photolysis, the biphenyl ($R1=R_2$=phenyl) substrate shows 20–30% label migration, implying 40–60% of product goes through the oxirene intermediate. α-diazocyclohexanone shows no label scrambling under photolytic conditions, as it is entirely s-*cis*, and thus all substrate goes through the concerted mechanism, avoiding the oxirene intermediate.

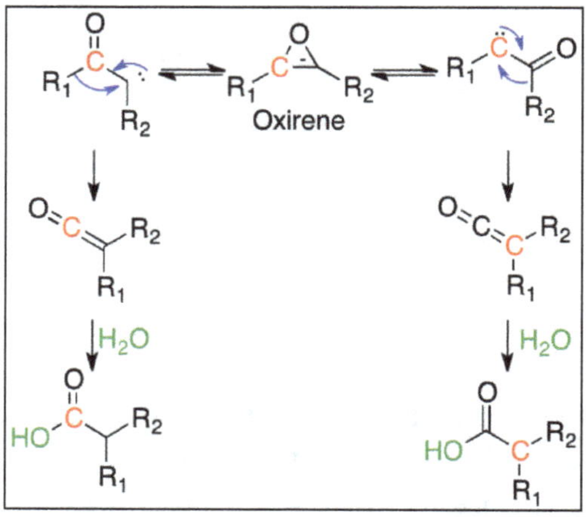


Isotopic labeling studies have been used extensively to measure the ratio of product stemming from a concerted mechanism versus a stepwise mechanism. These studies confirm that reactants that prefer s-*trans* conformations tend to undergo stepwise reaction. The degree of scrambling is also affected by carbene stability, migratory abilities, and nucleophilicity of solvent. The observation that the migratory ability of a substituent is inversely proportional to amount of carbene formed, indicates that under photolysis, there are competing pathways for many Wolff reactions. The only Wolff rearrangements that show no scrambling are s-*cis* constrained cyclic α-diazo ketones.

Under both thermolytic and photolytic conditions, there exist competing concerted and stepwise mechanisms. Many mechanistic studies have been carried out, including conformational, sensitization, kinetic, and isotopic scrambling studies. These all point to competing mechanisms, with general trends. α-Diazo ketones that exist in the s-*cis* conformation generally undergo a concerted mechanism, whereas those in the s-*trans* conformation undergo a stepwise mechanism. α-diazo ketones with better migratory groups prefer a concerted mechanism. However, for all substrates except cyclic α-diazo ketones that exist solely in the s-*cis* conformation, products come from a combination of both pathways. Transition metal mediated reactions are quite varied; however, they generally prefer forming the metal carbene intermediate. The complete mechanism under photolysis can be approximated in the following figure:

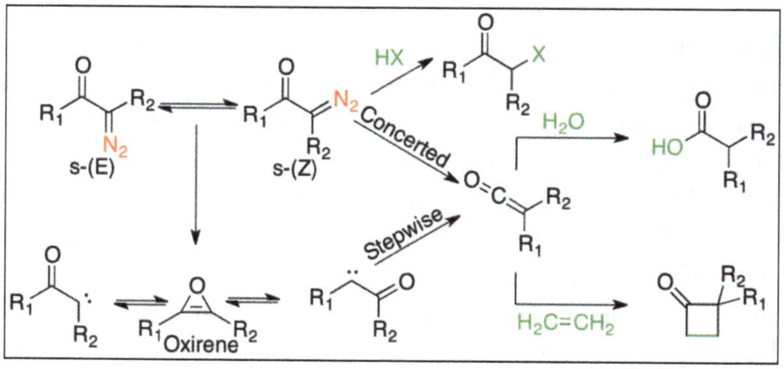

## Migratory Trends

The mechanism of the Wolff rearrangement is dependent on the aptitude of the migratory group. Migratory abilities have been determined by competition studies. In general, hydrogen migrates the fastest, and alkyl and aryl groups migrate at approximately the same rate, with alkyl migrations favored under photolysis, and aryl migrations preferred under thermolysis. Substituent effects on aryl groups are negligible, with the exception of $NO_2$, which is a poor migrator. In competition studies, electron deficient alkyl, aryl, and carbonyl groups cannot compete with other migrating groups, but are still competent. Heteroatoms, in general, are poor migratory groups, because their ability to donate electron density from their p orbitals into the π* C=O bond decreases migratory ability. The trend is as follows:

Photochemical reactions: H > alkyl ≥ aryl >> SR > OR ≥ NR2

Thermal reactions H > aryl ≥ alkyl (heteroatoms do not migrate)

## Preparation of α-diazocarbonyl Compounds

While known since 1902, the Wolff rearrangement did not become synthetically useful until the early 1930s, when efficient methods became available to synthesize α-diazocarbonyl compounds. The primary ways to prepare these substrates today are via the Arndt-Eistert procedure, the Franzen modification to the Dakin-West reaction, and diazo-transfer methods.

## Arndt-Eistert Procedure

The Arndt–Eistert reaction involves the acylation of diazomethane with an acid chloride, to yield a primary α-diazo ketone. The carbon terminus of diazomethane adds to the carbonyl, to create a tetrahedral intermediate, which eliminates chloride. The chloride then deprotonates the intermediate to give the α-diazo ketone product.

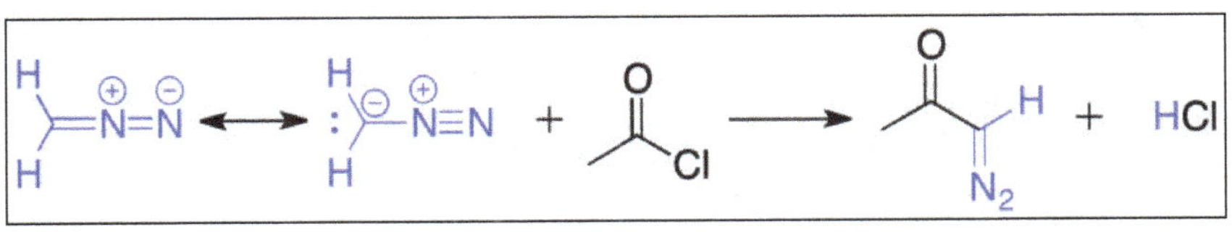

These α-diazo ketones are unstable under acidic conditions, as the α-carbon can be protonated by HCl and $S_N2$ displacement of nitrogen can occur by chloride.

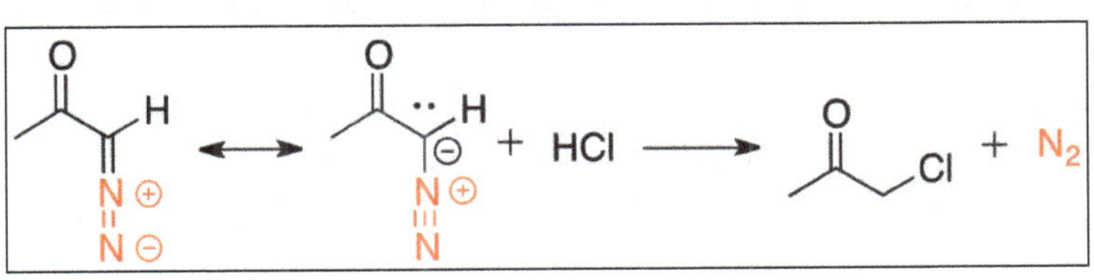

Thus, triethylamine or a second equivalent of diazomethane must be added to scavenge the HCl produced in order to prevent decomposition.

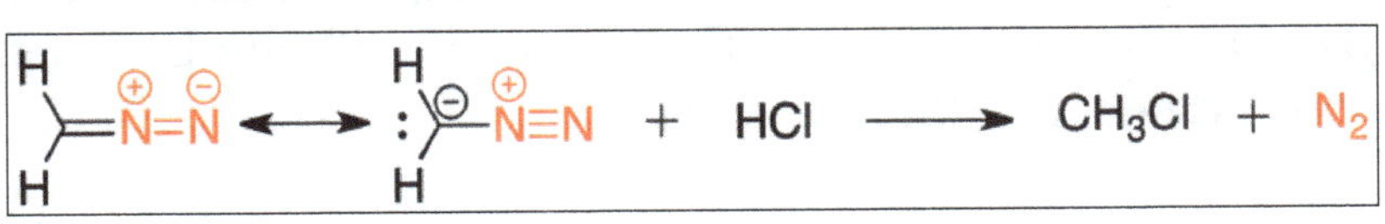

This reaction is occasionally done with acid anhydrides, which yields a one-to-one mixture of α-diazo ketone and the corresponding methyl ester. This method can also be used with primary diazoalkanes, to produce secondary α-diazo ketones. However, there are many limitations. Primary diazoalkanes undergo azo coupling to form azines; thus the reaction conditions must be altered such that acid chloride is added to a solution of diazoalkane and triethylamine at low temperature. In addition, primary diazoalkanes are very reactive, incompatible with acidic functionalities, and will react with activated alkenes including unsaturated carbonyls to give 1,3-dipolar cycloaddition products.

## Franzen Modification to Dakin-West Reaction

The Dakin–West reaction is a reaction of an amino acid with an acid anhydride in the presence of a base to form keto-amides. The Franzen modification to the Dakin–West reaction is a more effective way to make secondary α-diazo ketones. The Franzen modification nitrosates the keto-amide with $N_2O_3$ in acetic acid, and the resulting product reacts with methoxide in methanol to give the secondary α-diazo ketone.

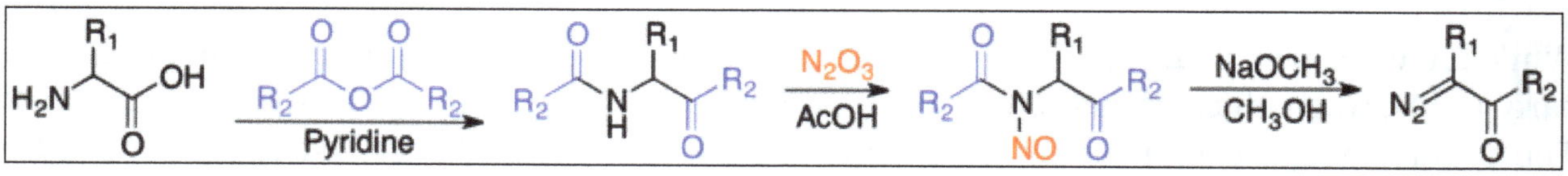

## Diazo-transfer Reactions

Diazo-transfer reactions are commonly used methods, in which an organic azide, usually tosylazide, and an activated methylene (i.e. a methylene with two withdrawing groups) react in the presence of a base to give an α-diazo-1,3-diketone. The base deprotonates the methylene, yielding an enolate, which reacts with tosylazide and subsequently decomposes in the presence of a weak acid, to give the α-diazo-1,3-diketone.

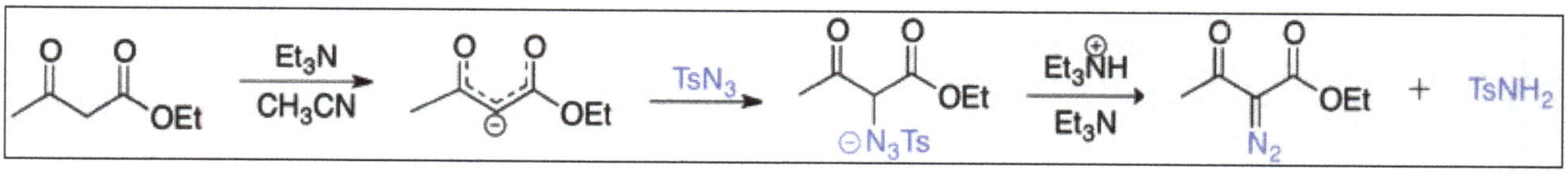

The necessary requirement of two electron withdrawing groups makes this reaction one of limited scope. The scope can be broadened to substrates containing one electron withdrawing group by formylating a ketone via a Claisen condensation, followed by diazo-transfer and deformylative group transfer.

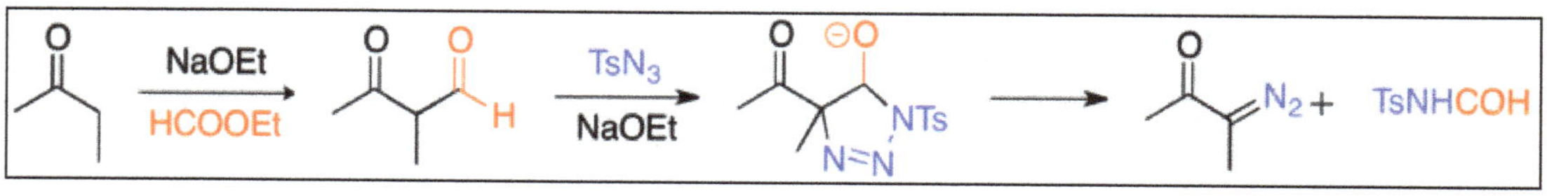

One of the greatest advantages of this method is its compatibility with unsaturated ketones. However, to achieve kinetic regioselectivity in enolate formation and greater compatibility with unsaturated carbonyls, one can induce enolate formation with lithium hexamethyldisilazide and subsequently trifluoroacylate rather than formylate.

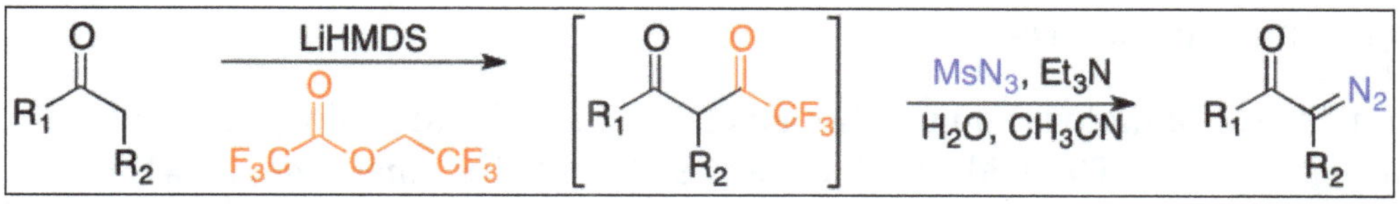


## Methods for Inducing Rearrangement

Wolff rearrangements can be induced under thermolytic, photolytic, and transition-metal-catalyzed conditions.

Thermal conditions to induce rearrangement require heating to relatively high temperatures, of 180 °C, and thus have limited use. Many Wolff rearrangement products are ring-strained and are susceptible to ring-open under high temperatures. In addition, $S_N2$ substitution of the diazo group at the α-carbon can take place at lower temperatures than rearrangement, which results in byproducts. The greatest use of thermal Wolff rearrangements is the formation of carboxylic acid analogs, by interception of the ketene with high boiling solvents, such as aniline and phenol.

Transition metals greatly lower the temperature of Wolff rearrangements, by stabilization of a metal-carbene intermediate. However, these carbenes can be so stable, as to not undergo rearrangement. Carbenes of rhodium, copper, and palladium are too stable and give non-Wolff products (primarily carbene insertion products). The most commonly used metal catalyst is silver(I) oxide, although silver benzoate is also common. These reactions are generally run in the presence of a weak base, such as sodium carbonate or tertiary amines.

Whereas thermal and metal mediated Wolff rearrangements date back to 1902, photolytic methods are somewhat newer, with the first example of a photolytic Wolff rearrangement reported in 1951. α-diazo ketones have two absorption bands, an allowed $\pi\rightarrow\pi^*$ transition at 240–270 nm, and a formally forbidden $\pi\rightarrow\sigma^*$ transition at 270–310 nm. Medium or low-pressure mercury arc lamps can excite these respective transitions. Triplet sensitizers result in non-Wolff carbene byproducts, and thus are not useful in synthetic applications of the Wolff rearrangement. However, they have been used to probe the mechanism of the Wolff rearrangement.

## Synthetic uses

The Wolff rearrangement has a few retrons, depending on the reaction out of the ketene intermediate. A carboxylic acid derivative with an α-methylene group is a retron for an Arndt-Eistert type homologation. An acid in which the α-carbon belongs to a ring is a retron for a Wolff rearrangement ring contraction.

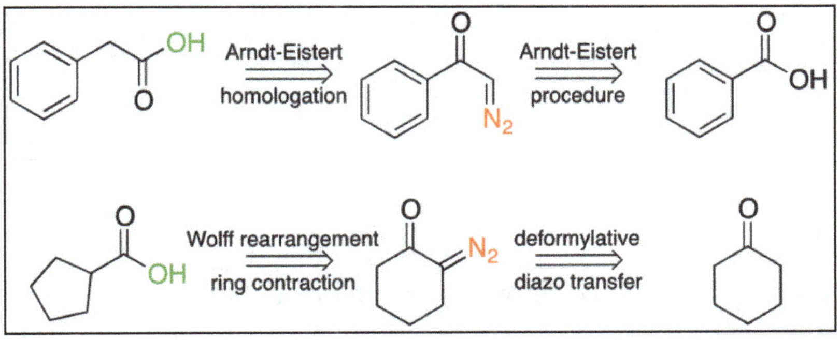

## Homologation Reactions

In the Arndt-Eistert homologation reaction, a carboxylic acid and thionyl chloride are reacted to generate an acid chloride. The acid chloride then reacts with diazomethane ($R_2$ = H), or occasionally a diazoalkyl, via the Arndt-Eistert procedure, to generate an α-diazo ketone, which will undergo a metal-catalyzed or photolyzed Wolff rearrangement, to give a ketene. The ketene can be trapped with any weak acid, such as an alcohol or amine, to form the ester or amide. However, trapping with water, to form the acid is the most common form.

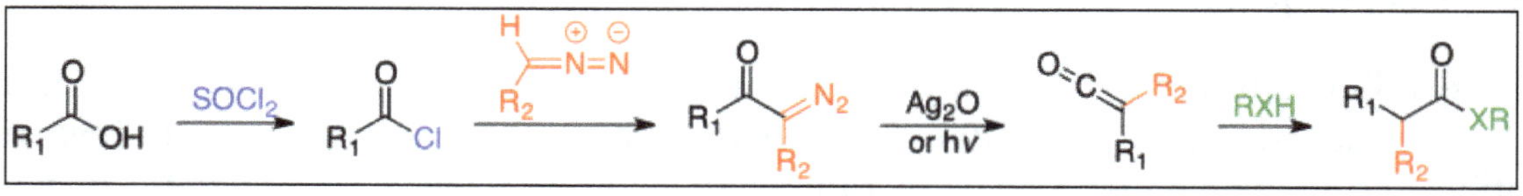


In the most basic form, where $R_2$= H, RXH=$H_2O$, the reaction lengthens the alkyl chain of a carboxylic acid by a methylene. However, there is great synthetic utility in the variety of reactions one can carry out, by varying the diazoalkyl and weak acid. The migrating group, $R_1$ migrates with complete retention. A very useful application of the Arndt-Eistert homologation forms the homologated aldehyde by either trapping the ketene with N-methyl aniline and reducing with lithium aluminum hydride, or trapping the ketene with ethanethiol and reducing with Raney nickel.

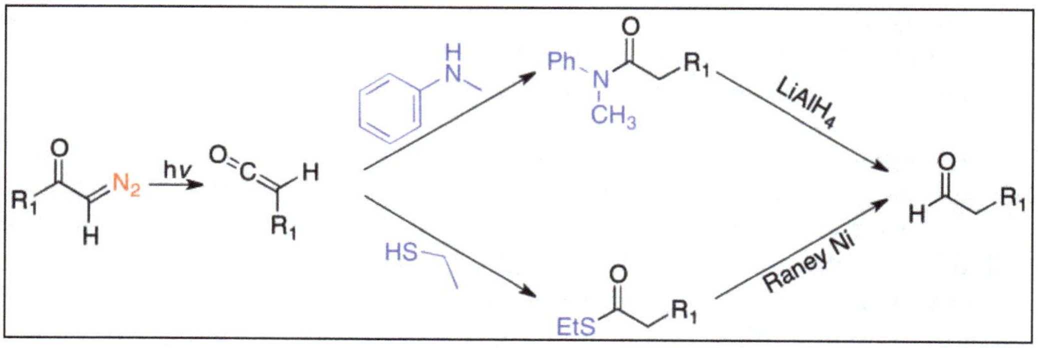


There exist many hundreds of examples of the Arndt-Eistert homologation in the literature. Prominent examples in natural product total synthesis include the syntheses of (–)-indolizidine and (+)-macbecin. A recent example of the Arndt-Eistert homologation is a step in the middle stage of Sarah Reisman's synthesis of (+)-salvileucalin B.

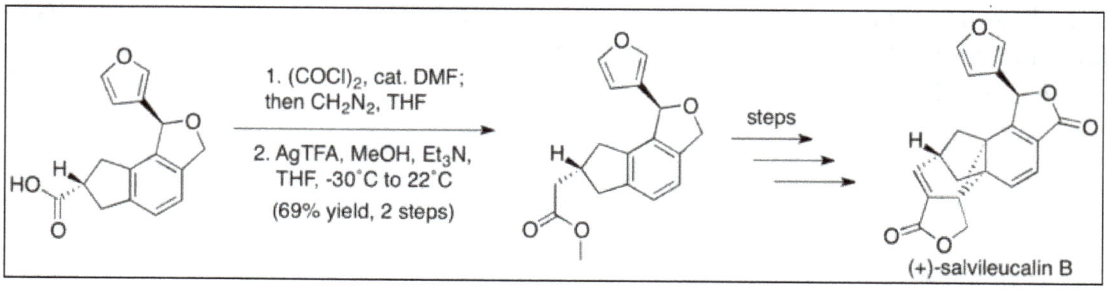


## Ring Contractions

If the reactant is a cyclic α-diazo ketone, then Wolff-rearrangement products will be the one-carbon ring-contracted product. These reactions are generally concerted due to the s-*cis* conformation, and are photocatalyzed. The reaction below shows the concerted mechanism for the ring contraction of α-diazocyclohexanone, followed by trapping of the ketene with a weakly acidic nucleophile.

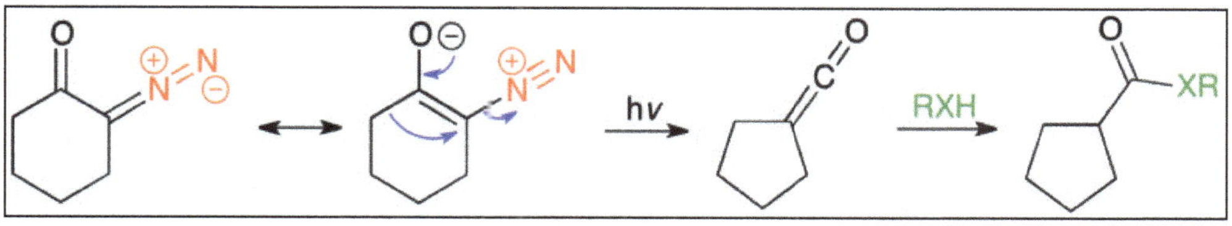


The first known example is the ring contracted Wolff rearrangement product of α-diazocamphor, and subsequent kinetic hydration of the ketene from the more sterically accessible "endo" face, to give *exo*-1,5,5-trimethylbicyclo hexane-6-carboxylic acid.

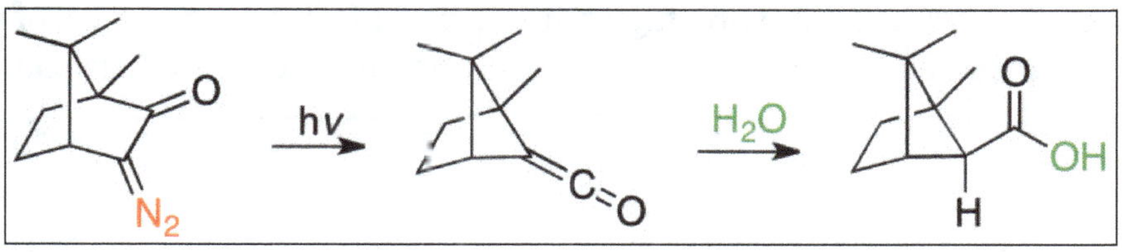


Ring contractions have been used extensively to build strained ring systems, as ring size does not impede the Wolff rearrangement, but often impedes other reactions. There are many examples where the Wolff rearrangement is used to contract cyclopentanone to cyclobutane. The rearrangement is commonly used to form strained bicyclic and ring-fused systems. There exist a handful of examples of ring contractions from cyclobutanones to cyclopropanes. The Wolff rearrangement is capable of contracting cyclohexanones to cyclopentanes, but is infrequently used to do so, because the Favorskii rearrangement accomplishes this transformation and the Wolff precursor is often more challenging to synthesize. However, an example of a cyclohexanone ring contraction using deformylative diazo transfer, followed by a Wolff rearrangement, is Keiichiro Fukumoto's synthesis of (±)-$\Delta^{9(12)}$-capnellene.

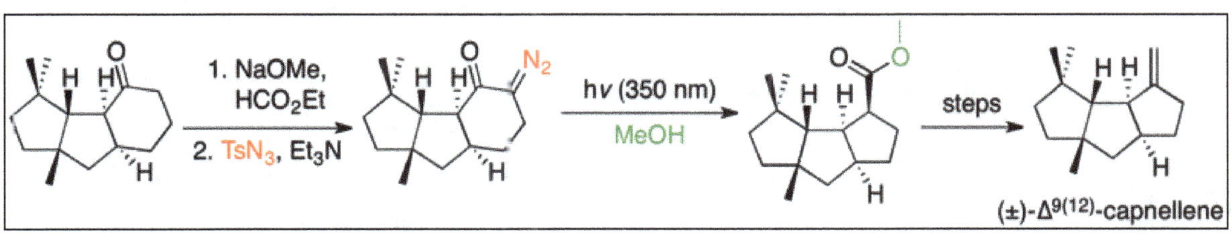


## Cycloaddition Reactions

Ketene intermediates produced via the Wolff rearrangement are well known to undergo [2 + 2] thermal cycloadditions with olefins to form four-membered rings in both intermolecular and intramolecular reactions, Ketenes are able to undergo what is normally considered a forbidden [2 + 2] cycloaddition reaction because the ketene acts in an antarrafactial manner, leading to the Woodward-Hoffmann allowed $[\pi_s^2 + \pi_\varepsilon^2]$ cycloaddition. Ketene [2 + 2] cycloadditions can be difficult reactions and give poor yields due to competing processes. The high energy aldoketene is very reactive and will cyclize with the diazo ketone starting material to produce butenolides and pyrazoles.

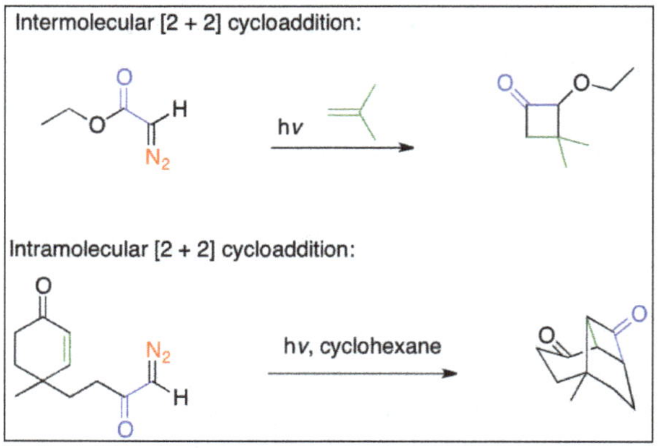


Ketene [2 + 2] cycloaddition reactions have been used in many total syntheses since Corey's use of the [2 + 2] cyclization in synthesizing the prostaglandins. Robert Ireland's synthesis of (±)-aphidicolin uses the Wolff rearrangement to do a tandem ring-contraction, and [2 + 2] cycloaddition.

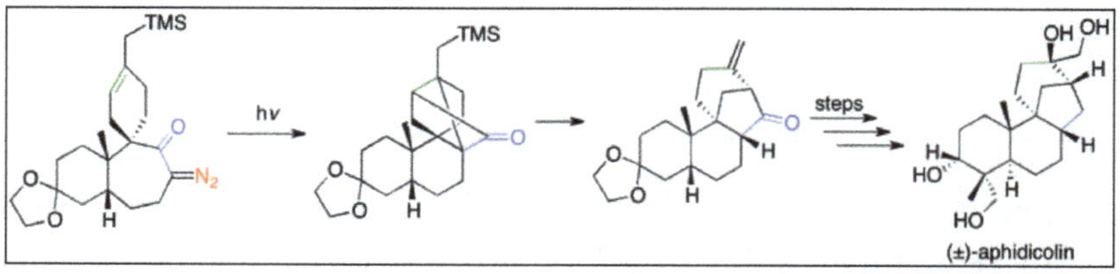


The Danheiser benzannulation photolyses α-diazo ketones and traps with an alkyne, which undergoes a pericyclic cascade, to ultimately form versatilely substituted phenols. The first step in the benzannulation is the photolysis of an α-diazo ketone to form a vinylketene. The vinylketene then undergoes a [2 + 2] cycloaddition with an alkyne to form a 2-vinylcyclobutenone, which does a 4π electrocyclic ring-opening to generate a dienylketene. The dienylketene subsequently undergoes a 6π electrocyclic ring-closure followed by tautomerization, to form the phenolic benzannulated product.

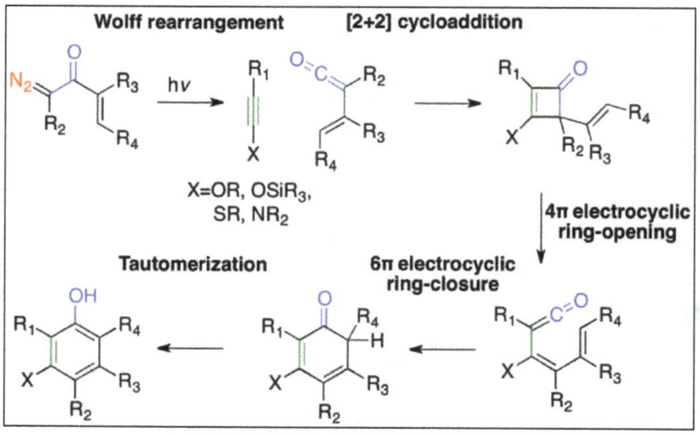


## Vinylogous Wolff Rearrangements

The vinylogous Wolff rearrangement consists of a β,γ-unsaturated diazo ketone undergoing a Wolff rearrangement, and a formal 1,3-shift of the $CH_2CO_2R$ group. The vinylogous Wolff rearrangement yields a γ,δ-unsaturated carboxylic acid derivative, which is the same retron as for the

Claisen rearrangement. The variant was discovered when it was noticed that thermolysis of 1-diazo-3,3,3-triarylpropan-2-ones gave unexpected isomeric products.

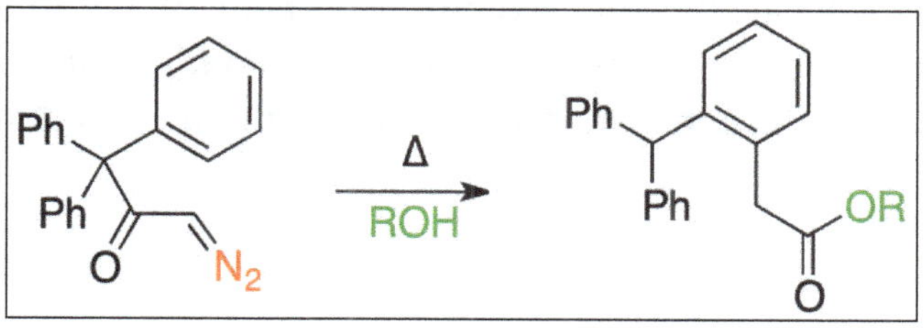


Copper (II) and rhodium (II) salts tend to give vinylogous Wolff rearranged products, and $CuSO_4$ and $Rh_2(OAc)_4$ are the most commonly used catalysts. This is because they promote metal carbene formation, which can add to the olefin to form a cyclopropane, which can reopen via a retro [2 + 2] to form a formally 1,3-shifted ketene (vis-à-vis a normal Wolff rearranged ketene), which can be trapped by a nucleophile to give the vinylogous Wolff product.

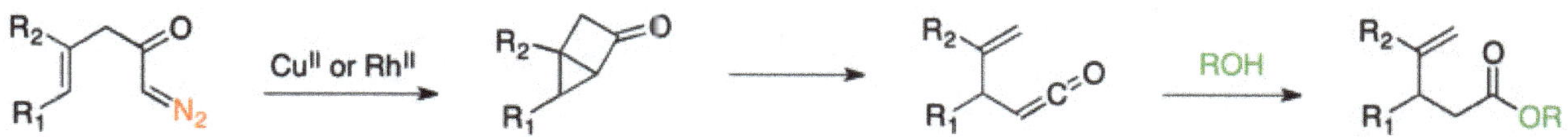


## Beckmann Rearrangement

The Beckmann rearrangement, named after the German chemist Ernst Otto Beckmann (1853–1923), is a rearrangement of an oxime functional group to substituted amides. The rearrangement has also been successful performed on haloimines and nitrones. Cyclic oximes and haloimines yield lactams.

The Beckmann rearrangement is often catalyzed by acid, however other reagents have been known to promote the rearrangement. These include tosyl chloride, thionyl chloride, phosphorus pentachloride, phosphorus pentoxide, triethylamine, sodium hydroxide, trimethylsilyl iodide among others. The Beckmann fragmentation is another reaction that often competes with the rearrangement, though careful selection of promoting reagent and solvent conditions can favor the formation of one over the other, sometimes giving almost exclusively one product. The rearrangement occurs stereospecifically for ketoximes and N-chloro/N-fluoro imines, with the migrating group being anti-periplanar to the leaving group on the nitrogen. Certain conditions have been known to racemize the oxime geometry, leading to the formation of both regioisomers. The rearrangement of aldoximes occurs with stereospecificity in the gas phase and without stereospecificity in the solution phase. A few methodologies allow for the rearrangement of aldoximes to primary amides, but fragmentation commonly competes in these systems. Nitrone rearrangement also occurs without stereospecificity; the regioisomer formed has the amide nitrogen substituted with the group possessing the greatest migratory aptitude.

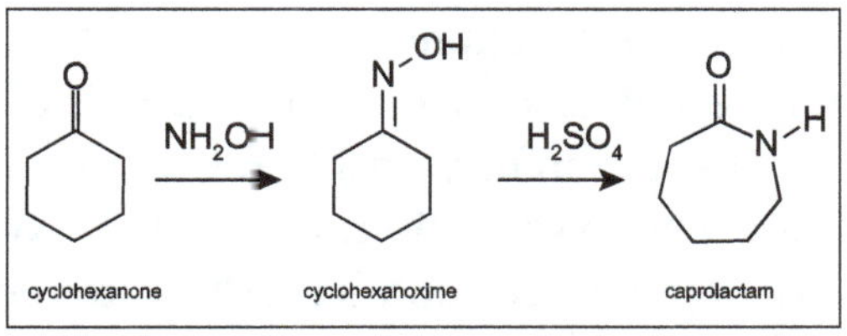


The archetypal Beckmann rearrangement is the conversion of cyclohexanone to caprolactam via the oxime. Caprolactam is the feedstock in the production of Nylon 6.

The Beckmann solution consists of acetic acid, hydrochloric acid and acetic anhydride, and was widely used to catalyze the rearrangement. Other acids, such as sulfuric acid, polyphosphoric acid, and hydrogen fluoride have all been used. Sulfuric acid is the most commonly used acid for commercial lactam production due to its formation of an ammonium sulfate by-product when neutralized with ammonia. Ammonium sulfate is a common agricultural fertilizer providing nitrogen and sulfur.

## Reaction Mechanism

The most common reaction mechanism of the Beckmann rearrangement consists generally of an alkyl migration anti-periplanar to the expulsion of a leaving group to form a nitrilium ion. This is followed by solvolysis to an imidate and then tautomerization to the amide:

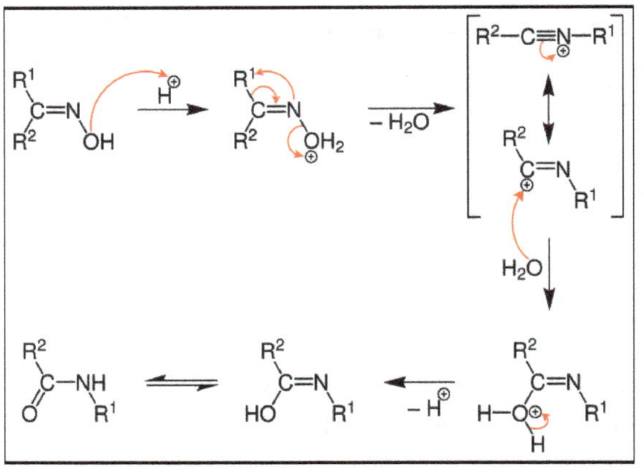

This nitrilium ion has been known to be intercepted by other nucleophiles, including the leaving group from the oxime.

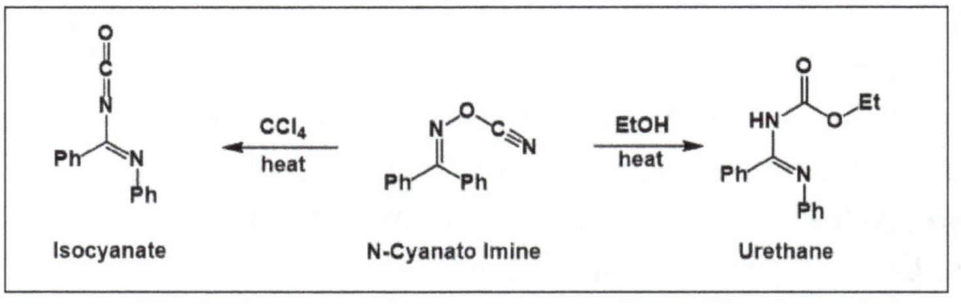


Presumably after the phenyl group migrates and expels the cyanate, it then attacks the nitrillium ion formed. In carbon tetrachloride the isocyanate can be isolated, whereas in ethanol the urethane is formed after solvolysis of the isocyanate.

One computational study has established the mechanism accounting for solvent molecules and substituents. The rearrangement of acetone oxime in the Beckmann solution involved three acetic acid molecules and one proton (present as an oxonium ion). In the transition state leading to the iminium ion (σ-complex), the methyl group migrates to the nitrogen atom in a concerted reaction as the hydroxyl group is expelled. The oxygen atom in the hydroxyl group is stabilized by three acetic acid molecules. In the next step the electrophilic carbon atom in the nitrilium ion is attacked by water and a proton is donated back to acetic acid. In the transition state leading to the imidate, the water oxygen atom is coordinated to 4 other atoms. In the third step, an isomerization step protonates the nitrogen atom leading to the amide.

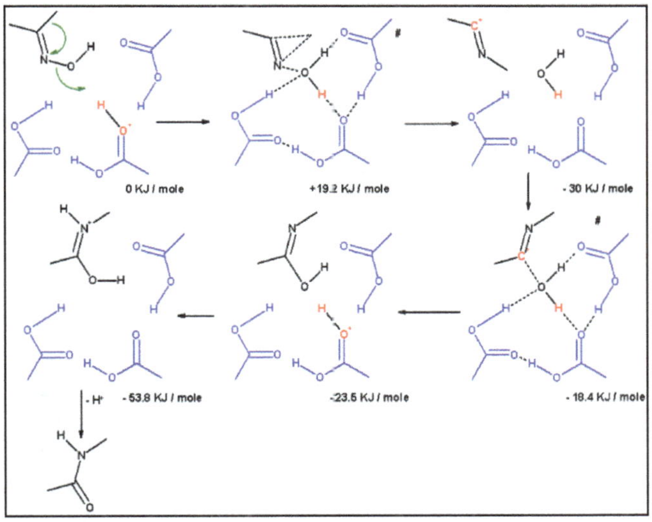

The same computation with a hydroxonium ion and 6 molecules of water has the same result, but when the migrating substituent is a phenyl group, the mechanism favors the formation of an intermediate three-membered π-complex. This π-complex is not found in the $H_3O^+(H_2O)_6$.

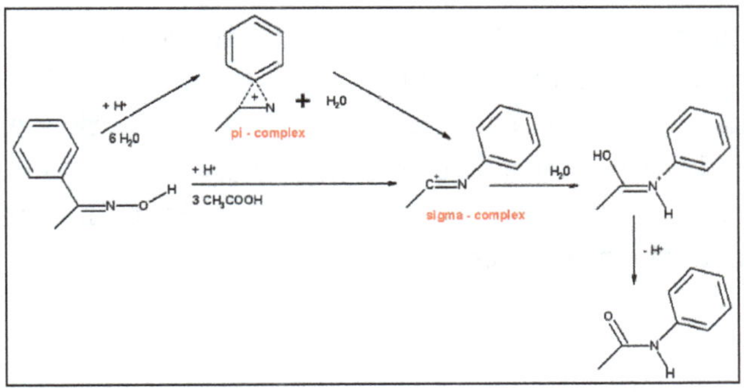

With the cyclohexanone-oxime, the relief of ring strain results in a third reaction mechanism, leading directly to the protonated caprolactam in a single concerted step without the intermediate formation of a π-complex or σ-complex.

## Cyanuric Chloride Assisted Beckmann Reaction

Beckmann rearrangement can be rendered catalytic using cyanuric chloride and zinc chloride as a co-catalyst. For example, cyclododecanone can be converted to the corresponding lactam, the monomer used in the production of Nylon 12.

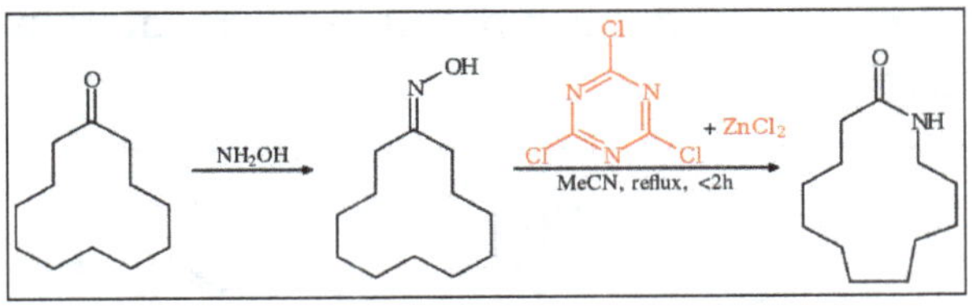

The reaction mechanism for this reaction is based on a catalytic cycle with cyanuric chloride activating the hydroxyl group via a nucleophilic aromatic substitution. The reaction product is dislodged and replaced by new reactant via an intermediate Meisenheimer complex.

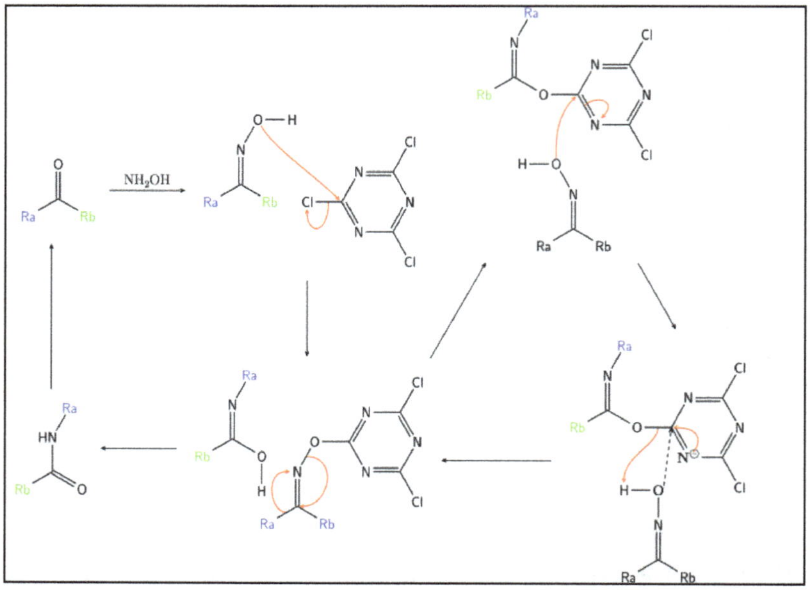

## Beckmann Fragmentation

The Beckmann fragmentation is a reaction that frequently competes with the Beckmann rearrangement. When the group α to the oxime is capable of stabilizing carbocation formation, the fragmentation becomes a viable reaction pathway. The reaction generates a nitrile and a carbocation, which is quickly intercepted to form a variety of products. The nitrile can also be hydrolyzed under reaction conditions to give carboxylic acids. Different reaction conditions can favor the fragmentation over the rearrangement.

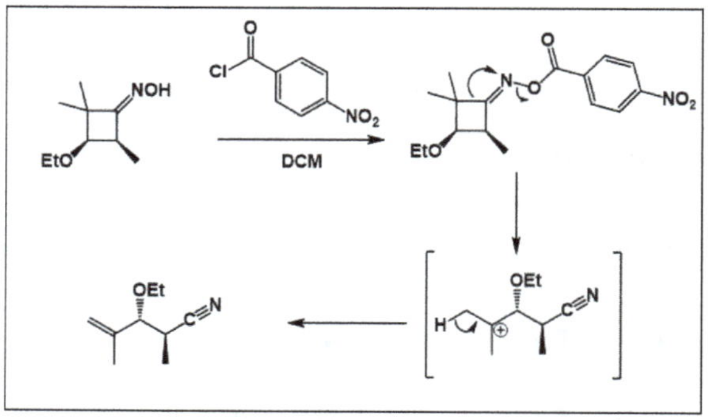

Quaternary carbon centers promote fragmentation by stabilizing carbocation formation through hyperconjugation. the "stable" carbocation is formed, which then loses a hydrogen to give a site of unsaturation. Oxygen and nitrogen atoms also promote fragmentation through the formation of ketones and imines respectively.

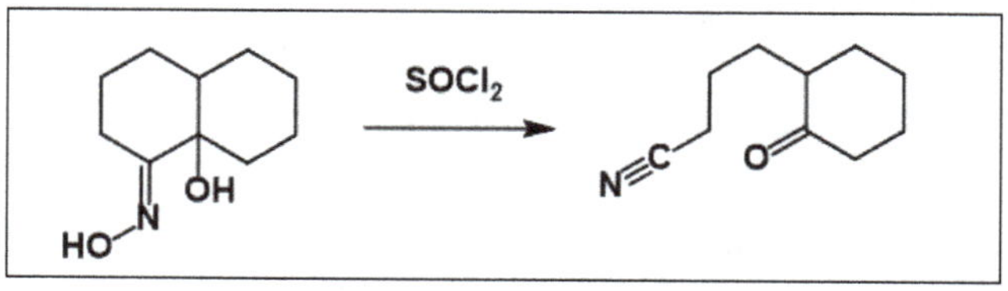

Sulfur is also capable of promoting fragmentation, albeit at a longer range than oxygen or nitrogen.

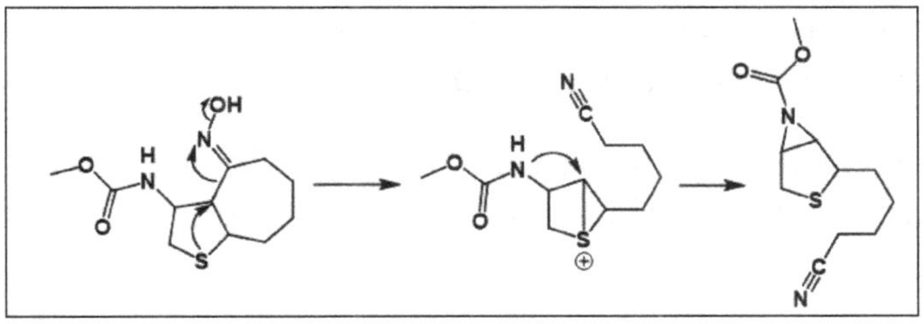

Silicon is capable of directing the fragmentation through the beta-silicon effect.

The carbocation intermediate in this reaction is intercepted by nucleophilic fluoride from diethylaminosulfur trifluoride (DAST):

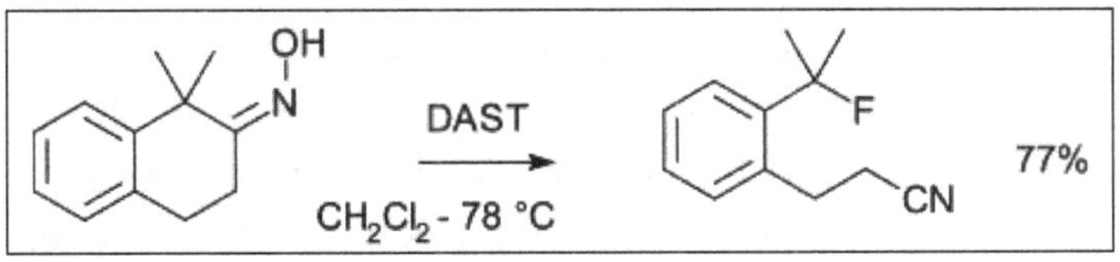


## Semmler–Wolff Reaction

The oxime of cyclohexenone with acid forms aniline in a dehydration – aromatization reaction called the Semmler–Wolff reaction or Wolff aromatization.

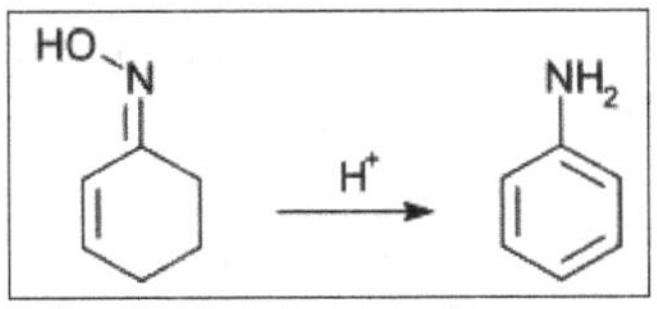


The mechanism can be shown as below:

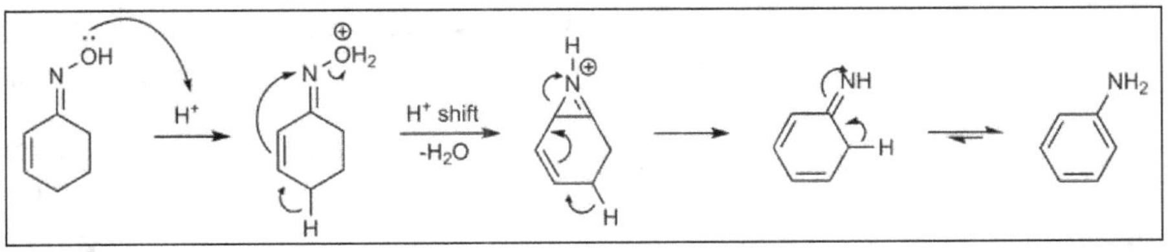


The reaction is intrinsically a special case of Beckmann rearrangement combined with neighbouring group participation.

## Applications in Drug Synthesis

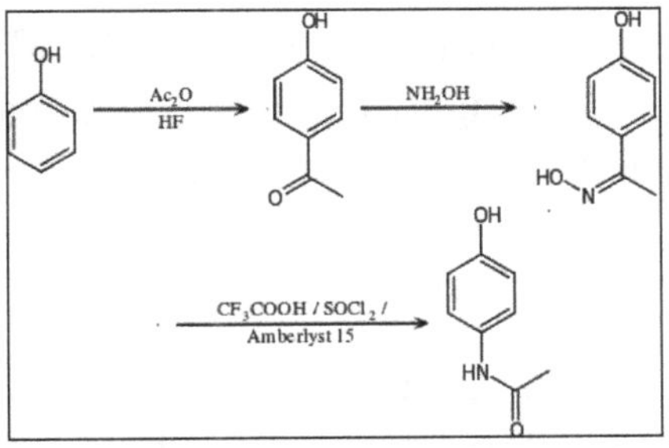

This route also involves the Beckmann rearrangement.

An industrial synthesis of paracetamol developed by Hoechst–Celanese involves the conversion of a methyl ketone to an acetanilide via a Beckmann rearrangement.

The Beckmann rearrangement is also used in the synthesis of DHEA, benazepril, ceforanide, elanzepine, 17-azaprogesterone, elantrine, prazepine, enprazepine, and etazepine.

## Oxidation-reduction Reactions

Oxidation-reduction reactions are of central importance in organic chemistry. In ionic and free radical reactions, oxidation and reduction are defined as processes by which an element undergoes a net loss or gain of electrons, respectively. The concept as applied to organic covalent compounds, where elements share electrons rather than losing or gaining them is the same, but it's frequently simplified and narrowed down to make it easier to recognize these processes. Therefore it must be kept in mind that, while the following definition is grossly simplified, it serves the goal of quickly identifying oxidation and reduction processes in many organic reactions.

In reference to organic molecules, oxidation is a process by which a carbon atom gains bonds to more electronegative elements, most commonly oxygen. Reduction is a process by which a carbon atom gains bonds to less electronegative elements, most commonly hydrogen.

The following chart summarizes these concepts when applied to organic transformations. [ox] stands for oxidation, and [red] stands for reduction.

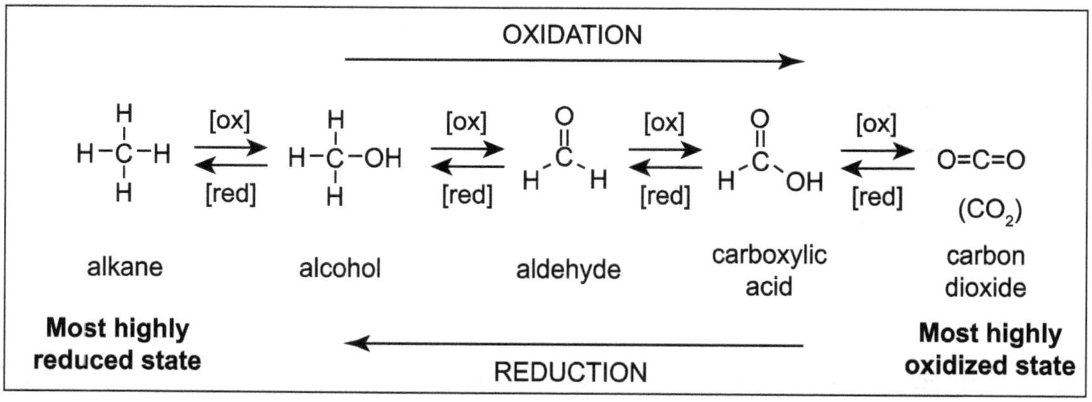


Some points must be noted. First, a gain or loss of bonds means simply more or less bonds. The double bond counts as two bonds and the triple bond counts as three bonds. Thus in the carbonyl group (C=O) carbon is considered to have two bonds to oxygen. Therefore, this carbon has a higher oxidation state than the alcohol carbon, which has only one bond to oxygen. In routine terminology, it is said that the aldehyde is a more highly oxidized functional group than the alcohol. Oxidation reactions are therefore those in which the central carbon of a functional group is transformed into a more highly oxidized form, and reduction reactions are those in which the central carbon is transformed into a more highly reduced form.

Second, there can be several functional groups where the central carbon has the same oxidation state. For example, the carbons bonded to oxygen in alcohols and ethers have the same oxidation state. Likewise for aldehydes and their hydrated forms, and for carboxylic acids and their derivatives. However, most references to oxidation and reduction reactions in organic chemistry textbooks involve the functional groups presented in the chart above.

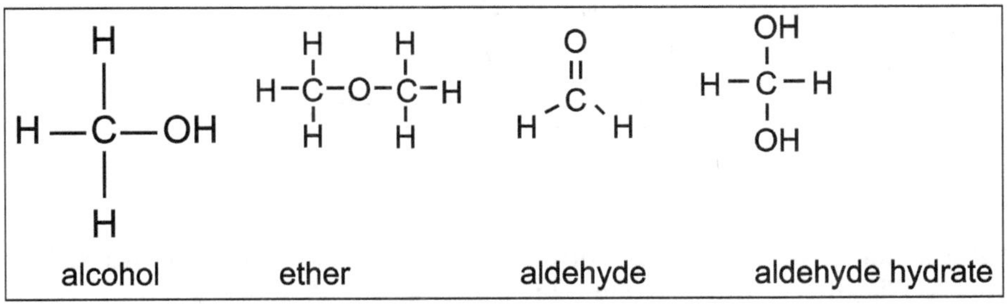


The chart presented before shows the oxidation and reduction states for a molecule that contains only one carbon. But most organic compounds contain more than one carbon. The maximum oxidation state that a particular carbon can attain depends on how many other carbons it must remain attached to. For example, a molecule with two carbon atoms could not be oxidized all the way to carbon dioxide because the carbon atom in $CO_2$ must have four bonds to oxygen, leaving no room for bonds to other carbons. The maximum oxidation state that a carbon can attain decreases gradually as the number of bonds to other carbons increases. Thus, the maximum oxidation state possible for a carbon that's bonded to one other carbon is the carboxylic acid stage, and so on. The following chart illustrates this idea.

### One Single Carbon

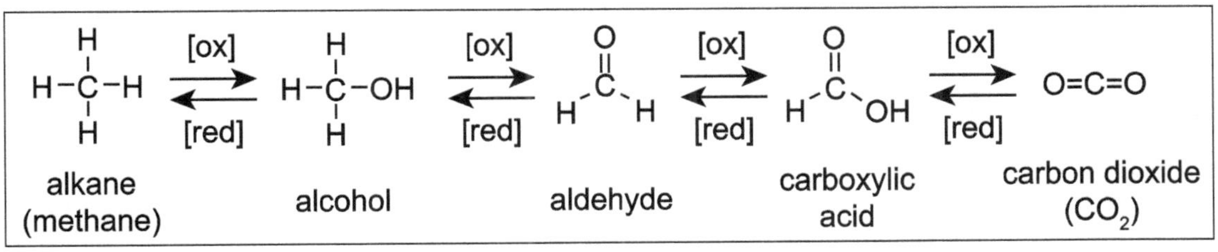


### Carbon Bonded to One Other Carbon

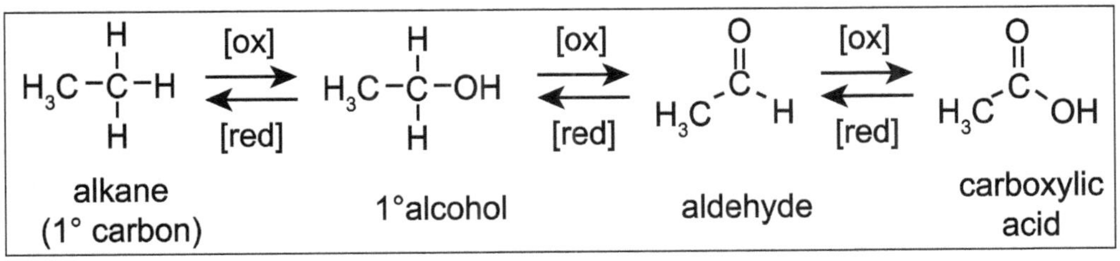


### Carbon Bonded to Two Other Carbons

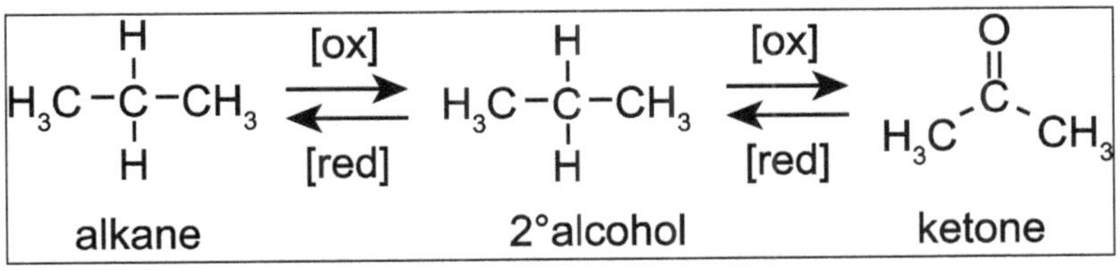

## Carbon Bonded to Three Other Carbons

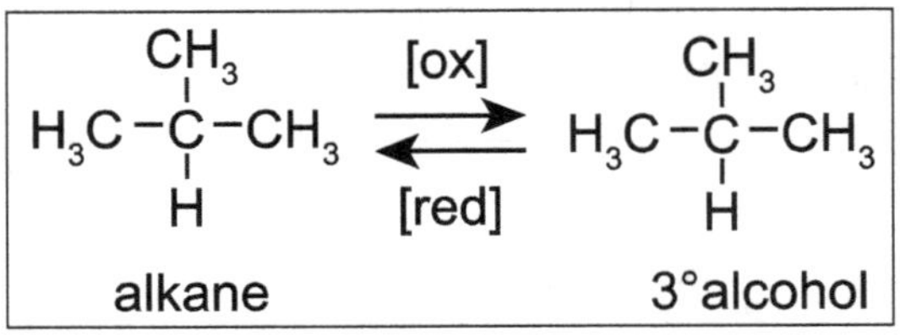


## References


- Smith, Michael (2007). March's advanced organic chemistry reactions, mechanisms, and structure (6th ed.). Hoboken, N.J.: Wiley-Interscience. pp. 1488–1493. ISBN 978-1-61583-842-4
- Lukeman, Matthew; Scaiano, Juan C. (2005). "Carbanion-Mediated Photocages: Rapid and Efficient Photorelease with Aqueous Compatibility". Journal of the American Chemical Society. 127 (21): 7698–7699. doi:10.1021/ja0517062. ISSN 0002-7863. PMID 15913358
- Types-of-organic-reactions, chemistry: byjus.com, Retrieved 10 June, 2019
- Plagens, Andreas; Laue, Thomas M. (2005). Named organic reactions (2nd ed.). Chichester: John Wiley & Sons. ISBN 0-470-01041-X
- Lang, S.; Murphy, J. A. (2006). "Azide rearrangements in electron-deficient systems". Chem. Soc. Rev. 35 (2): 146–156. doi:10.1039/B505080D. PMID 16444296

# 4

# Inorganic Compounds

A compound which lacks a carbon-hydrogen bond is generally known as an inorganic compound. Some of the common inorganic compounds are ammonium bicarbonate, sodium fluoride, beryllium oxide, calcium sulfate and sodium amide. The topics elaborated in this chapter will help in gaining a better perspective about these different inorganic compounds.

An inorganic compound is any compound that lacks a carbon atom, for lack of a more in-depth definition. Those compounds with a carbon atom are called organic compounds, due to their root base in an atom that is vital for life. There are a small number of inorganic compounds that actually do contain carbon, given its propensity for forming molecular bonds; these include carbon monoxide and carbon dioxide, to name a few.

Inorganic compounds are often quite simple, as they do not form the complex molecular bonds that carbon makes possible. A common example of a simple inorganic compound would be sodium chloride, known more commonly as household salt. This compound contains only two atoms, sodium (Na) and chlorine (Cl).

## Examples of Inorganic Compounds

1. $H_2O$ - Water is a simple inorganic compound, even though it contains hydrogen, a key atom (along with carbon) in many organic compounds. The atoms in a molecule of water have formed very simple bonds due to this lack of carbon.

2. HCl - Hydrochloride, also known as hydrochloric acid when it is dissolved in water, is a colorless, corrosive acid with a fairly strong pH. It is found in the gastric juices of many animals, helping in digestion by breaking down food.

3. $CO_2$ - Carbon dioxide, despite the presence of a carbon atom in the formula, is classified as an inorganic compound. This has caused a dispute within the scientific community, with questions being raised as to the validity of our current methods of classifying compounds. Currently, organic compounds contain a carbon or a hydrocarbon, which forms a stronger bond. The bond formed by carbon in $CO_2$ is not a strong bond.

4. $NO_2$ - Nitrogen dioxide gas presents a variety of colors at different temperatures. It is often produced in atmospheric nuclear tests, and is responsible for the tell-tale reddish color displayed in mushroom clouds. It is highly toxic, and forms fairly weak bonds between the nitrogen and oxygen atoms.

5. $Fe_2O_3$ - Iron (III) oxide is one of the three main oxides of iron, and is an inorganic compound due to the lack of a carbon atom or a hydrocarbon. Iron (III) oxide occurs naturally as hematite, and is the source of most iron for the steel production industry. It is commonly known as rust, and shares a number of characteristics with its naturally ocurring counterpart.

## The Main Classes of Inorganic Compounds

Simple substances. Molecules consist of one-type atoms (atoms of one element). In chemical reactions cannot be decomposed with forming other substances.

Complex substances (or chemical compounds). Molecules consist of different type atoms (atoms of different chemical elements). In chemical reactions are decomposed with formation several other substances.

| Inorganic substances | |
|---|---|
| Simple | Metals |
| | Non-metals |
| Complex | Oxides |
| | Bases |
| | Acids |
| | Salts |

Sharp border between metals and non-metals does not exist, since be simple substances showing dual properties.

### Bases

Bases - complex substances, in which atoms of metals bonded with one or several hydroxyls groups (according to electrolytic dissociation theory bases - complex substances, which under the dissociating in water solution are formed metal cations (or $NH_4^+$) and hydroxide anions $OH^-$).

## Classification

Soluble in water (alkalis) and insoluble. Amphoteric bases show also properties of weak acids.

### Preparation

1. Reactions of active metals (alkaline and alkaline earth metals) with water

$$2Na + 2H_2O \rightarrow 2NaOH + H_2$$

$$Ca + 2H_2O \rightarrow Ca(OH)_2 + H_2$$

2. Interaction oxides of active metals with water

$$BaO + H_2O \rightarrow Ba(OH)_2$$

3. Electrolysis water solutions of salts

$$2NaCl + 2H_2O \rightarrow 2NaOH + H_2 + Cl_2$$

## Chemical Properties

| Alkalis | Insoluble bases |
|---|---|
| 1. Action to indicators | |
| litmus - blue<br>methylorange - yellow<br>phenolphthalein - crimson | -- |
| 2. Interaction with acid oxides | |
| $2KOH + CO_2 \rightarrow K_2CO_3 + H_2O$<br>$KOH + CO_2 \rightarrow KHCO_3$ | -- |
| 3. Interaction with acids (reaction of neutralization) | |
| $NaOH + HNO_3 \rightarrow NaNO_3 + H_2O$ | $Cu(OH)_2 + 2HCl \rightarrow CuCl_2 + 2H_2O$ |
| 4. Reaction of exchange with salts | |
| $Ba(OH)_2 + K_2SO_4 \rightarrow 2KOH + BaSO_4\downarrow$<br>$3KOH + Fe(NO_3)_3 \rightarrow Fe(OH)_3I + 3KNO_3$ | -- |
| 5. Thermal decomposition | |
| -- | $Cu(OH)_2 \xrightarrow{t^o} CuO + H_2O$ |

## Oxides

## Classification

Oxides - complex substances, consisting from two elements, one of which oxygen.

| Oxides | |
|---|---|
| Non-salts forming<br>$(CO, N_2O, NO)$ | |
| Salts forming | Basic-<br>it is a metal oxides in which metals display<br>low oxidation number +1, +2<br>$Na_2O$; $MgO$; $CuO$ |
| | Amphoteric<br>(for metals with oxidation number +3, +4).<br>As a hydrates it corresponding amphoteric hydroxide<br>$ZnO$; $Al_2O_3$; $Cr_2O_3$; $SnO_2$ |
| | Acid-<br>it is oxide of nonmetals and metals with oxidation<br>number from +5 to +7.<br>$SO_2$; $SO_3$; $P_2O_5$; $Mn_2O_7$; $CrO_3$ |

Amphoteric basic corresponding bases; amphoteric acid- corresponding acids.

## Preparation

1. Interaction of simple and complex substances with oxygen:

$$2Mg + O_2 \rightarrow 2MgO$$

$$4P + 5O_2 \rightarrow 2P_2O_5$$

$$S + O_2 \rightarrow SO_2$$

$$2CO + O_2 \rightarrow 2CO_2$$

$$2CuS + 3O_2 \rightarrow 2CuO + 2SO_2$$

$$CH_4 + 2O_2 \rightarrow CO_2 + 2H_2O$$

$$4NH_3 + 5O_2 \xrightarrow[(s)]{cat.} 4NO + 6H_2O$$

2. Decomposition some substances containing oxygen (bases, acids, salts) under the heating:

$$Cu(OH)2 \xrightarrow{t^\circ} CuO + H_2O$$

$$(CuOH)_2CO_3 \xrightarrow{t^\circ} 2CuO + CO_2 + H_2O$$

$$2Pb(NO_3)_2 \xrightarrow{t^\circ} 2PbO + 4NO_2 + O$$

$$2HMnO_4 \xrightarrow[t^\circ]{H_2SO_4\,(conc.)} Mn_2O_7 + H_2O$$

## Chemical Properties

| Basic oxides | Acid oxides |
|---|---|
| 1. Interaction with water | |
| Base formed:<br>$Na_2O + H_2O \rightarrow 2NaOH$<br>$CaO + H_2O \rightarrow Ca(OH)_2$ | Acid formed:<br>$SO_3 + H_2O \rightarrow H_2SO_4$<br>$P_2O_5 + 3H_2O \rightarrow 2H_3PO_4$ |
| 2. Interactionwithacidorbase | |
| On reactions with acid salt and water are formed<br>$MgO + H_2SO_4 \xrightarrow{t^\circ} MgSO_4 + H_2O$<br>$CuO + 2HCl \xrightarrow{t^\circ} CuCl_2 + H_2O$ | On reactions with base salt and water are formed<br>$CO_2 + Ba(OH)_2 \rightarrow BaCO_3 + H_2O$<br>$SO_2 + 2NaOH \rightarrow Na_2SO_3 + H_2O$ |
| Amphoteric oxides interact | |
| with acids as basic:<br>$ZnO + H_2SO_4 \rightarrow ZnSO_4 + H_2O$ | with bases as acid:<br>$ZnO + 2NaOH \rightarrow Na_2ZnO_2 + H_2O$<br>$(ZnO + 2NaOH + H_2O \rightarrow Na_2[Zn(OH)_4])$ |

| 3. Interaction of basic and acid oxide with each other leads to salt formation |
|---|
| $Na_2O + CO_2 \rightarrow Na_2CO_3$ |
| 4. Reduction up to simple substances: |
| $3CuO + 2NH_3 \rightarrow 3Cu + N_2 + 3H_2O$<br>$P_2O_5 + 5C \rightarrow 2P + 5CO$ |

## Acids

Acids - complex substances, consisting from hydrogen atoms and acid radical (according to electrolytic dissociation theory: acids - electrolytes, which under the dissociating form only $H^+$ in the capacity of cations).

## Classification

1. On composition: oxygenless and oxoacids.
2. On hydrogen atoms number, which capable to substituted on metal: mono-, di-, tribasic.

| Oxygenless: | | Salt's name: |
|---|---|---|
| hCl - Hydrogen chloride (hydrochloric) | Monobasic | Chloride |
| HBr - Hydrogen bromide | Monobasic | Bromide |
| hI - Hydrogeniodide | Monobasic | Iodide |
| HF - Hydrogen fluorine (hydrofluoric) | Monobasic | Fluoride |
| $h_2S$ - Hydrogensulphide | Bibasic | Sulphide |
| Containing oxygen: | | |
| $HNO_3$ – Nitric | Monobasic | Nitrate |
| $H_2SO_3$ – Sulphurous | Bibasic | Sulphite |
| $H_2SO_4$ – Sulphuric | Bibasic | Sulphate |
| $H_2CO_3$ – Carbonic | Bibasic | Carbonate |
| $H_2SiO_3$ – Silicon | Bibasic | Silicate |
| $H_3PO_4$ – Ortophosphoric | Tribasic | Ortophosphate |

## Preparation

1. Interaction of acid oxides with water (for oxoacids):

$$SO_3 + H_2O \rightarrow H_2SO_4$$

$$P_2O_5 + 3H_2O \rightarrow 2H_3PO_4$$

2. Interaction of hydrogen with non-metals and following dissolution product in water (for oxygenless acids).

$$H_2 + Cl_2 \rightarrow 2HCl$$

$$H_2 + S \rightarrow H_2S$$

3. Reactions of exchange between salt and acid

$$Ba(NO_3)_2 + H_2SO_4 \rightarrow BaSO_4\downarrow + 2HNO_3$$

including displacement weak, flying or slightly soluble acid from its salts by means of more strong acids.

$$Na_2SiO_3 + 2HCl \rightarrow H_2SiO_3\downarrow + 2NaCl$$

$$2NaCl\,(hard) + H_2SO_4\,(conc.) \xrightarrow{t^\circ} Na_2SO_4 + 2HCl$$

## Chemical Properties

1. Action to indicators.

litmus - red

methylorange – pink

2. Interaction with bases (reaction of neutralisation)

$$H_2SO_4 + 2KOH \rightarrow K_2SO_4 + 2H_2O$$

$$2HNO_3 + Ca(OH)_2 \rightarrow Ca(NO_3)_2 + 2H_2O$$

3. Interaction with basic oxides.

$$CuO + 2HNO_3 \xrightarrow{t^\circ} Cu(NO_3)_2 + H_2O$$

4. Interaction with metals.

$$Zn + 2HCl \rightarrow ZnCl_2 + H_2$$

$$2Al + 6HCl \rightarrow 2AlCl_3 + 3H_2$$

(metals standing in the electrochemical series before hydrogen, acid-oxidizers).

5. Interaction with salts (reactions of exchange) at which stands out gas or formed residual.

$$H_2SO_4 + BaCl_2 \rightarrow BaSO_4\downarrow + 2HCl$$

$$2HCl + K_2CO_3 \rightarrow 2KCl + H_2O + CO_2$$

## Salts

Salts - complex substances, which consist from atoms of metal and acid residuals. This the most multiple class of inorganic compounds.

## Classification

| Salts | Medium |
|---|---|
| | Acid |
| | Basic |
| | Double |
| | Mixed |
| | Complex |

Medium salts. In the time of dissociation give only metal cations (or $NH_4^+$) and anions of acid radical. Products of full substitution hydrogen atoms of acids to atoms of metals:

$$Na_2SO_4 \leftrightarrow 2Na^+ + SO_4^{2-}$$

$$CaCl_2 \leftrightarrow Ca^{2+} + 2Cl^-$$

Acid salts. In the time of dissociation give only metal cations (or $NH_4^+$), hydrogen anions and anions of acid radical. Products of full substitution hydrogen atoms of multibasic acid to atoms of metal:

$$NaHCO_3 \leftrightarrow Na^+ + HCO_3^- \leftrightarrow Na^+ + H^+ + CO_3^{2-}$$

Basic salts. In the time of dissociation give only metal cations, hydroxyl anions and anions of acid radical. Products of incomplete substitution OH groups, corresponding bases to acid radicals:

$$Zn(OH)Cl \leftrightarrow [Zn(OH)]^+ + Cl^- \leftrightarrow Zn^{2+} + OH^- + Cl^-$$

Double salts. In the time of dissociation gives two cations and one anion:

$$KAl(SO_4)_2 \leftrightarrow K^+ + Al^{3+} + 2SO_4^{2-}$$

Mixed salts. Formed by means of one cation and two anions:

$$CaOCl_2 \leftrightarrow Ca^{2+} + Cl^- + OCl^-$$

Complex salts. Contain complex cations and anions:

$$[Ag(NH_3)_2]Br \leftrightarrow [Ag(NH_3)_2]^+ + Br^-$$

$$Na[Ag(CN)_2] \leftrightarrow Na^+ + [Ag(CN)_2]^-$$

## Medium Salts

## Preparation

Most of ways of getting the salts is based on the interaction of substances with opposite properties:

1. Metal with non-metal:

$$2Na + Cl_2 \rightarrow 2NaCl$$

2. Metal with acid:

$$Zn + 2HCl \rightarrow ZnCl_2 + H_2$$

3. Metal with solution of salt of less active metal:

$$Fe + CuSO_4 \rightarrow FeSO_4 + Cu$$

4. Basic oxide with the acid oxide:

$$MgO + CO_2 \rightarrow MgCO_3$$

5. Basic oxide with acid:

$$CuO + H_2SO_4 \xrightarrow{t^\circ} CuSO_4 + H_2O$$

6. Bases with acid oxide:

$$Ba(OH)_2 + CO_2 \rightarrow BaCO_3\downarrow + H_2O$$

7. Bases with acid:

$$Ca(OH)_2 + 2HCl \rightarrow CaCl_2 + 2H_2O$$

8. Salts with the acid:

$$MgCO_3 + 2HCl \rightarrow MgCl_2 + H_2O + CO_2$$

$$BaCl_2 + H_2SO_4 \rightarrow BaSO_4I + 2HCl$$

9. Bases solution with salt solution:

$$Ba(OH)_2 + Na_2SO_4 \rightarrow 2NaOH + BaSO_4I$$

10. Solutions of two salts:

$$3CaCl_2 + 2Na_3PO_4 \rightarrow Ca_3(PO_4)_2I + 6NaCl$$

## Chemical Properties

1. Thermal decomposition

$$CaCO_3 \rightarrow CaO + CO_2$$

$$2Cu(NO_3)_2 \rightarrow 2CuO + 4NO_2 + O_2$$

$$NH_4Cl \rightarrow NH_3 + HCl$$

2. Hydrolysis

$$Al_2S_3 + 6H_2O \leftrightarrow 2Al(OH)_3I + 3H_2S$$

$$FeCl_3 + H_2O \leftrightarrow Fe(OH)Cl_2 + HCl$$

$$Na_2S + H_2O \leftrightarrow NaHS + NaOH$$

3. Exchange reactions with acids, bases and other salts

$$AgNO_3 + HCl \rightarrow AgCl\downarrow + HNO_3$$

$$Fe(NO_3)_3 + 3NaOH \downarrow Fe(OH)_3\downarrow + 3NaNO_3$$

$$CaCl_2 + Na_2SiO_3 \rightarrow CaSiO_3\downarrow + 2NaCl$$

4. Oxidation-reduction reactions, stipulated by properties of cation or anion

$$2KMnO_4 + 16HCl \rightarrow 2MnCl_2 + 2KCl + 5Cl_2 + 8H_2O$$

## Acid Salts

## Preparation

1. Interaction of acid with the deficit of basis.

$$KOH + H_2SO_4 \rightarrow KHSO_4 + H_2O$$

2. Interaction of bases with plenty acid oxides.

$$Ca(OH)_2 + 2CO_2 \rightarrow Ca(HCO_3)_2$$

3. Interaction of medium salts with acid.

$$Ca_3(PO_4)_2 + 4H_3PO_4 \rightarrow 3Ca(H_2PO_4)_2$$

## Chemical Properties

1. Thermal decomposition with medium salts formation.

$$Ca(HCO_3)_2 \rightarrow CaCO_3\downarrow + CO_2 + H_2O$$

2. Interaction with the alkali. Reception of medium salts.

$$Ba(HCO_3)_2 + Ba(OH)_2 \rightarrow 2BaCO_3I + 2H_2O$$

## Basic Salts

## Preparation

1. Hydrolysis of salts, formed by weak base and strong acid.

$$ZnCl_2 + H_2O \rightarrow [Zn(OH)]Cl + HCl$$

2. Addition (by drops) a small quantities of alkalis to solutions of medium salts of metals.

$$AlCl_3 + 2NaOH \rightarrow [Al(OH)_2]Cl + 2NaCl$$

3. Interaction of weak acids salts with medium salts.

$$2MgCl_2 + 2Na_2CO_3 + H_2O \rightarrow [Mg(OH)]_2CO_3I + CO_2 + 4NaCl$$

## Chemical Properties

1. Thermal decomposition.

$$[Cu(OH)]_2CO_3 \rightarrow 2CuO + CO_2 + H_2O$$
malachite

2. Interaction with the acid: formation of medium salts.

$$Sn(OH)Cl + HCl \rightarrow SnCl_2 + H_2O$$

## Complex Salts

## Structure

| $K_4[Fe(CN)_6]$ | |
|---|---|
| $K_4[Fe(CN)_6]$ | – External spher |
| $K_4[Fe(CN)_6]$ | – Internal sphere |
| $K_4[Fe(CN)_6]$ | – Central atom |
| $K_4[Fe(CN)_6]$ | – Coordinate relation |
| $K_4[Fe(CN)_6]$ | – Ligand |

Central, complex forming, atoms is usually serves ions of metals of greater periods (Co, Ni, Pt, Hg, Ag, Cu). Typical ligands are $OH^-$, $CN^-$, $NH_3$, CO, $H_2O$; they connected with the central atom by donor-acceptor bound.

## Preparation

Reactions of salts with ligands:

$$AgCl + 2NH_3 \rightarrow [Ag(NH_3)_2]Cl$$

$$FeCl_3 + 6KCN \rightarrow K_3[Fe(CN)_6] + 3KCl$$

## Chemical Properties

1. Destruction of complexes at the expense of forming the slightly-soluble compounds:

$$2[Cu(NH_3)_2]Cl + K_2S \rightarrow CuSI + 2KCl + 4NH_3$$

2. The exchange of ligands between external and internal spheres:

$$K_2[CoCl_4] + 6H_2O \rightarrow [Co(H_2O)_6]Cl_2 + 2KCl$$

# Ammonium Bicarbonate

Ammonium bicarbonate is an inorganic compound with formula $(NH_4)HCO_3$, simplified to $NH_5CO_3$. The compound has many names, reflecting its long history. Chemically speaking, it is the bicarbonate salt of the ammonium ion. It is a colourless solid that degrades readily to carbon dioxide, water and ammonia.

## Production

Ammonium bicarbonate is produced by combining carbon dioxide and ammonia:

$$CO_2 + NH_3 + H_2O \rightarrow (NH_4)HCO_3$$

Since ammonium bicarbonate is thermally unstable, the reaction solution is kept cold, which allows the precipitation of the product as white solid. About 100,000 tons were produced in this way in 1997.

Ammonia gas passed into a strong aqueous solution of the sesquicarbonate (a 2:1:1 mixture of $(NH_4)HCO_3$, $(NH_4)_2CO_3$, and $H_2O$) converts it into normal ammonium carbonate ($(NH_4)_2CO_3$), which can be obtained in the crystalline condition from a solution prepared at about 30 °C. This compound on exposure to air gives off ammonia and reverts to ammonium bicarbonate.

## Salt of Hartshorn

Compositions containing ammonium carbonate have long been known. They were once produced commercially, formerly known as sal volatile or salt of hartshorn. It was obtained by the dry distillation of nitrogenous organic matter such as hair, horn, leather. In addition to ammonium bicarbonate, this material contains ammonium carbamate ($NH_4CO_2NH_2$), and ammonium carbonate ($(NH_4)_2CO_3$). It is sometimes called ammonium sesquicarbonate. It possesses a strong ammoniacal smell, and on digestion with alcohol, the carbamate is dissolved leaving a residue of ammonium bicarbonate.

A similar decomposition takes place when the sesquicarbonate is exposed to air.

## Uses

Ammonium bicarbonate is used in the food industry as a raising agent for flat baked goods, such as cookies and crackers, and in China in steamed buns and Chinese almond cookies. It was commonly used in the home before modern-day baking powder was made available. In China it is called edible or food-grade "smelly powder". Many baking cookbooks (especially from Scandinavian countries) may still refer to it as hartshorn or hornsalt (e.g., FI: "hirvensarvisuola", NO: "hjortetakksalt", DK: "hjortetakssalt", SE: "hjorthornssalt", "salt of hart's horn", DE: "Hirschhornsalz"). Although there is a slight smell of ammonia during baking, this quickly dissipates, leaving no taste. It is used in, for example, Swedish "drömmar" biscuits and Danish Christmas biscuits (Hjortetakssalt), and German Lebkuchen. In many cases it may be substituted with baking soda or baking powder or a combination of both, depending on the recipe composition and leavening requirements. Compared to baking soda or potash, hartshorn has the advantage of producing more gas for the same amount of agent, and of not leaving any salty or soapy taste in the finished product, as it completely decomposes into water and gaseous products that evaporate during baking. It cannot be used for moist, bulky baked goods however, such as normal bread or cakes, since some ammonia will be trapped inside and will cause an unpleasant taste.

It is commonly used as an inexpensive nitrogen fertilizer in China, but is now being phased out in favor of urea for quality and stability. This compound is used as a component in the production of fire-extinguishing compounds, pharmaceuticals, dyes, pigments, and it is also a basic fertilizer being a source of ammonia. Ammonium bicarbonate is still widely used in the plastic and rubber industry, in the manufacture of ceramics, in chrome leather tanning, and for the synthesis of catalysts.

It is also used for buffering solutions to slightly alkaline pH during chemical purification, such as high-performance liquid chromatography. Because it entirely decomposes to volatile compounds this allows rapid recovery of the compound of interest by freeze-drying.

Ammonium bicarbonate is also a key component of the expectorant cough syrup "Senega and Ammonia".

E-Number: E503

### Reactions

It dissolves in water to give a mildly alkaline solution. It is insoluble in acetone and alcohols.

Ammonium bicarbonate decomposes above about 36 °C into ammonia, carbon dioxide, and water in an endothermic process and so causes a drop in the temperature of the water:

$$NH_4HCO_3 \rightarrow NH_3 + H_2O + CO_2.$$

When treated with acids, carbon dioxide is also produced:

$$NH_4HCO_3 + HCl \rightarrow NH_4Cl + CO_2 + H_2O.$$

Reaction with base produces ammonia.

It reacts with sulfates of alkaline-earth metals to precipitating their carbonates:

$$CaSO_4 + 2\,NH_4HCO_3 \rightarrow CaCO_3 + (NH_4)_2SO_4 + CO_2 + H_2O.$$

It also reacts with alkali metal halides, giving alkali metal bicarbonate and ammonium halide:

$$NH_4HCO_3 + NaCl \rightarrow NH_4Cl + NaHCO_3;$$

$$NH_4HCO_3 + KI \rightarrow NH_4I + KHCO_3;$$

$$NH_4HCO_3 + NaBr \rightarrow NH_4Br + NaHCO_3.$$

## Ammonium Chloride

Ammonium chloride is an inorganic compound with the formula $NH_4Cl$ and a white crystalline salt that is highly soluble in water. Solutions of ammonium chloride are mildly acidic. Sal ammoniac is a name of the natural, mineralogical form of ammonium chloride. The mineral is commonly formed on burning coal dumps from condensation of coal-derived gases. It is also found around some types of volcanic vents. It is mainly used as fertilizer and a flavouring agent in some types of liquorice. It is the product from the reaction of hydrochloric acid and ammonia.

## Production

It is a product of the Solvay process used to produce sodium carbonate:

$$CO_2 + 2\,NH_3 + 2\,NaCl + H_2O \rightarrow 2\,NH_4Cl + Na_2CO_3$$

In addition to being the principal method for the manufacture of ammonium chloride, that method is used to minimize ammonia release in some industrial operations.

Ammonium chloride is prepared commercially by combining ammonia ($NH_3$) with either hydrogen chloride (gas) or hydrochloric acid (water solution):

$$NH_3 + HCl \rightarrow NH_4Cl$$

Ammonium chloride occurs naturally in volcanic regions, forming on volcanic rocks near fume-releasing vents (fumaroles). The crystals deposit directly from the gaseous state and tend to be short-lived, as they dissolve easily in water.

## Reactions

Ammonium chloride appears to sublime upon heating but actually decomposes into ammonia and hydrogen chloride gas.

$$NH_4Cl \rightarrow NH_3 + HCl$$

Ammonium chloride reacts with a strong base, like sodium hydroxide, to release ammonia gas:

$$NH_4Cl + NaOH \rightarrow NH_3 + NaCl + H_2O$$

Similarly, ammonium chloride also reacts with alkali metal carbonates at elevated temperatures, giving ammonia and alkali metal chloride:

$$2\,NH_4Cl + Na_2CO_3 \rightarrow 2\,NaCl + CO_2 + H_2O + 2\,NH_3$$

A 5% by weight solution of ammonium chloride in water has a pH in the range 4.6 to 6.0.

Some of ammonium chloride's reactions with other chemicals are endothermic like its reaction with barium hydroxide and its dissolving in water.

## Applications

The dominant application of ammonium chloride is as a nitrogen source in fertilizers (corresponding to 90% of the world production of ammonium chloride) such as chloroammonium phosphate. The main crops fertilized this way are rice and wheat in Asia.

Ammonium chloride crystal(s).

Ammonium chloride was used in pyrotechnics in the 18th century but was superseded by safer and less hygroscopic chemicals. Its purpose was to provide a chlorine donor to enhance the green and blue colours from copper ions in the flame.

It had a secondary use to provide white smoke, but its ready double decomposition reaction with potassium chlorate producing the highly unstable ammonium chlorate made its use very dangerous.

### Metalwork

Ammonium chloride is used as a flux in preparing metals to be tin coated, galvanized or soldered. It works as a flux by cleaning the surface of workpieces by reacting with the metal oxides at the surface to form a volatile metal chloride. For that purpose, it is sold in blocks at hardware stores for use in cleaning the tip of a soldering iron, and it can also be included in solder as flux.

### Medicine

Ammonium chloride is used as an expectorant in cough medicine. Its expectorant action is caused by irritative action on the bronchial mucosa, which causes the production of excess respiratory tract fluid, which presumably is easier to cough up. Ammonium salts are an irritant to the gastric mucosa and may induce nausea and vomiting.

Ammonium chloride is used as a systemic acidifying agent in treatment of severe metabolic alkalosis, in oral acid loading test to diagnose distal renal tubular acidosis, to maintain the urine at an acid pH in the treatment of some urinary-tract disorders.

### Food

Ammonium chloride, under the name *sal ammoniac* or *salmiak* is used as food additive under the E number E510, working as a yeast nutrient in breadmaking and as an acidifier. It is a feed supplement for cattle and an ingredient in nutritive media for yeasts and many microorganisms.

Ammonium chloride is used to spice up dark sweets called salmiak (popular in Nordic and other nearby countries), in baking to give cookies a very crisp texture, and in the liquor Salmiakki Koskenkorva for flavouring. In Iran, Tajikistan, India, Pakistan and Arab countries it is called "Noshader" and is used to improve the crispness of snacks such as *samosas* and *jalebi*.

### In the Laboratory

Ammonium chloride has been used historically to produce low temperatures in cooling baths. Ammonium chloride solutions with ammonia are used as buffer solutions including ACK (Ammonium-Chloride-Potassium) lysis buffer.

In paleontology, ammonium chloride vapor is deposited on fossils, where the substance forms a brilliant white, easily removed and fairly harmless and inert layer of tiny crystals. That covers up any coloration the fossil may have, and if lighted at an angle highly enhances contrast in photographic documentation of three-dimensional specimens. The same technique is applied in archaeology to eliminate reflection on glass and similar specimens for photography. In organic synthesis saturated $NH_4Cl$ solution is typically used to quench reaction mixtures.

### Floatation

Giant squid and some other large squid species maintain neutral buoyancy in seawater through an ammonium chloride solution which is found throughout their bodies and is less dense than seawater. This differs from the method of flotation used by most fish, which involves a gas-filled swim bladder.

### Other Applications

Ammonium chloride is used in a ~5% aqueous solution to work on oil wells with clay swelling problems. It is also used as electrolyte in zinc–carbon batteries. Other uses include in hair shampoo, in the glue that bonds plywood, and in cleaning products. In hair shampoo, it is used as a thickening agent in ammonium-based surfactant systems such as ammonium lauryl sulfate. Ammonium chloride is used in the textile and leather industry, in dyeing, tanning, textile printing and cotton clustering. Around the turn of the 20th century, ammonium chloride was used in aqueous solution as the electrolyte in Leclanché cells. These cells found a commercial use as the "local battery" in subscribers' telephone installations. In woodworking, a solution of ammonium chloride and water, when applied to unfinished wood, will burn when subjected to a heat gun resulting in a branding iron mark without use of a branding iron. The solution can be painted onto the wood or applied with a common rubber stamp.

## Antimony Trioxide

Antimony(III) oxide is the inorganic compound with the formula $Sb_2O_3$. It is the most important commercial compound of antimony. It is found in nature as the minerals valentinite and senarmontite. Like most polymeric oxides, $Sb_2O_3$ dissolves in aqueous solutions with hydrolysis.

### Production and Properties

Global production of antimony(III) oxide in 2012 was 130,000 tonnes, an increase from 112,600 tonnes in 2002. China produces the largest share followed by US/Mexico, Europe, Japan and South Africa and other countries (2%).

As of 2010, antimony(III) oxide was produced at four sites in EU27. It is produced via two routes, re-volatilizing of crude antimony(III) oxide and by oxidation of antimony metal. Oxidation of antimony metal dominates in Europe. Several processes for the production of crude antimony(III) oxide or metallic antimony from virgin material. The choice of process depends on the composition of the ore and other factors. Typical steps include mining, crushing and grinding of ore, sometimes followed by froth flotation and separation of the metal using pyrometallurgical processes (smelting or roasting) or in a few cases (e.g. when the ore is rich in precious metals) by hydrometallurgical processes. These steps do not take place in the EU but closer to the mining location.

## Re-volatilizing of Crude Antimony(III) Oxide

Crude stibnite is oxidized to crude antimony(III) oxide using furnaces operating at approximately 500 to 1,000 °C. The reaction is the following:

$$2\ Sb_2S_3 + 9\ O_2 \rightarrow 2\ Sb_2O_3 + 6\ SO_2$$

The crude antimony(III) oxide is purified by sublimation.

## Oxidation of Antimony Metal

Antimony metal is oxidized to antimony(III) oxide in furnaces. The reaction is exothermic. Antimony(III) oxide is formed through sublimation and recovered in bag filters. The size of the formed particles is controlled by process conditions in furnace and gas flow. The reaction can be schematically described by:

$$4\ Sb + 3\ O_2 \rightarrow 2\ Sb_2O_3$$

## Properties

Antimony(III) oxide is an amphoteric oxide, it dissolves in aqueous sodium hydroxide solution to give the meta-antimonite $NaSbO_2$, which can be isolated as the trihydrate. Antimony(III) oxide also dissolves in concentrated mineral acids to give the corresponding salts, which hydrolyzes upon dilution with water. With nitric acid, the trioxide is oxidized to antimony(V) oxide.

When heated with carbon, the oxide is reduced to antimony metal. With other reducing agents such as sodium borohydride or lithium aluminium hydride, the unstable and very toxic gas stibine is produced. When heated with potassium bitartrate, a complex salt potassium antimony tartrate, $KSb(OH)_2{\cdot}C_4H_2O_6$ is formed.

## Structure

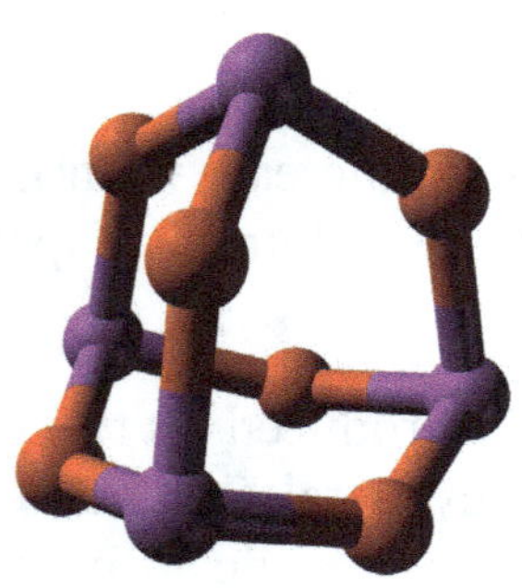

$Sb_4O_6$

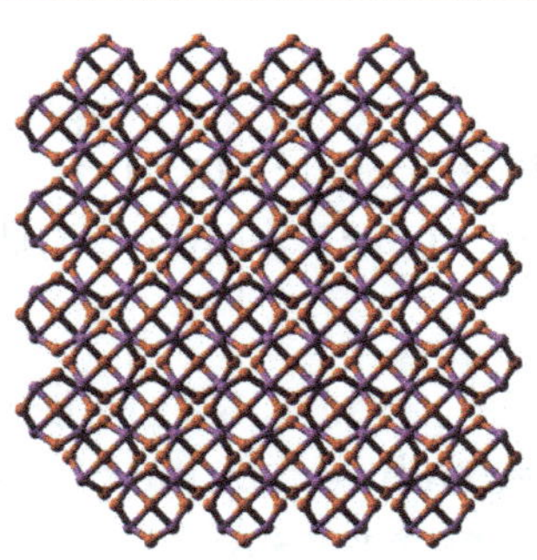

senarmontite

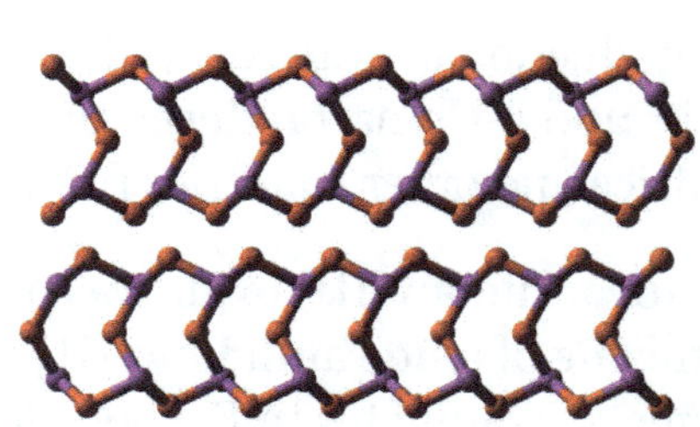

valentinite

The structure of $Sb_2O_3$ depends on the temperature of the sample. Dimeric $Sb_4O_6$ is the high temperature (1560 °C) gas. $Sb_4O_6$ molecules are bicyclic cages, similar to the related oxide of phosphorus(III), phosphorus trioxide. The cage structure is retained in a solid that crystallizes in a cubic habit. The Sb-O distance is 197.7 pm and the O-Sb-O angle of 95.6°. This form exists in nature as the mineral senarmontite. Above 606 °C, the more stable form is orthorhombic, consisting of pairs of -Sb-O-Sb-O- chains that are linked by oxide bridges between the Sb centers. This form exists in nature as the mineral valentinite.

## Uses

The annual consumption of antimony(III) oxide in the United States and Europe is approximately 10,000 and 25,000 tonnes, respectively. The main application is as flame retardant synergist in combination with halogenated materials. The combination of the halides and the antimony is key to the flame-retardant action for polymers, helping to form less flammable chars. Such flame retardants are found in electrical apparatuses, textiles, leather, and coatings.

Other applications:

- Antimony(III) oxide is an opacifying agent for glasses, ceramics and enamels.
- Some specialty pigments contain antimony.
- Antimony(III) oxide is a useful catalyst in the production of polyethylene terephthalate (PET plastic) and the vulcanization of rubber.

# Magnesium Hydroxide

Magnesium hydroxide is the inorganic compound with the chemical formula Mg(OH)2. It occurs in nature as the mineral brucite. It is a white solid with low solubility in water (Ksp = 5.61×10−12). Magnesium hydroxide is a common component of antacids, such as milk of magnesia, as well as laxatives.

## Preparation

Combining a solution of many magnesium salts with alkaline water induces precipitation of solid $Mg(OH)_2$:

$$Mg^{2+} + 2\ OH^- \rightarrow Mg(OH)_2$$

On a commercial scale, $Mg(OH)_2$ is produced by treating seawater with lime ($Ca(OH)_2$). 600 m³ of seawater gives about one ton of $Mg(OH)_2$. $Ca(OH)_2$ is far more soluble than $Mg(OH)_2$, so the latter precipitates as a solid:

$$MgO + H_2O \longrightarrow Mg(OH)_2$$

## Uses

## Precursor to MgO

Most $Mg(OH)_2$ that is produced industrially, as well as the small amount that is mined, is converted to fused magnesia (MgO). Magnesia is valuable because it is both a poor electrical conductor and an excellent thermal conductor.

## Health

## Metabolism

Magnesium hydroxide is used in suspension as either an antacid or a laxative, depending on concentration.

As an antacid, magnesium hydroxide is dosed at approximately 0.5–1.5 g in adults and works by simple neutralization, where the hydroxide ions from the $Mg(OH)_2$ combine with acidic $H^+$ ions produced in the form of hydrochloric acid by parietal cells in the stomach to produce water.

As a laxative, magnesium hydroxide is dosed at 2–5 g, and works in a number of ways. First, $Mg^{2+}$ is poorly absorbed from the intestinal tract, so it draws water from the surrounding tissue by osmosis. Not only does this increase in water content soften the feces, it also increases the volume of feces in the intestine (intraluminal volume) which naturally stimulates intestinal motility. Furthermore, $Mg^{2+}$ ions cause the release of cholecystokinin (CCK), which results in intraluminal accumulation of water, electrolytes, and increased intestinal motility. Although it has been stated in some sources, the hydroxide ions themselves do not play a significant role in the laxative effects of milk of magnesia, as basic solutions (i.e., solutions of hydroxide ions) are not strongly laxative, and non-basic $Mg^{2+}$ solutions, like $MgSO_4$, are equally strong laxatives mole for mole.

Only a small amount of the magnesium from magnesium hydroxide is usually absorbed by the intestine (unless one is deficient in magnesium). However, magnesium is mainly excreted by the kidneys so long-term, daily consumption of milk of magnesia by someone suffering from kidney failure could lead in theory to hypermagnesemia. Unabsorbed drug is excreted in feces; absorbed drug is excreted rapidly in urine.

## Milk of Magnesia

On May 4, 1818, American inventor John Callen received a patent (No. X2952) for magnesium hydroxide. In 1829, Sir James Murray used a "condensed solution of fluid magnesia" preparation of his own design to treat the Lord Lieutenant of Ireland, the Marquis of Anglesey, of stomach pain. This was so successful (advertised in Australia and approved by the Royal College of Surgeons in 1838) that he was appointed resident physician to Anglesey and two subsequent Lords Lieutenant, and knighted. His fluid magnesia product was patented two years after his death in 1873.

The term *milk of magnesia* was first used by Charles Henry Phillips in 1872 for a suspension of magnesium hydroxide formulated at about 8%w/v. It was sold under the brand name *Phillips' Milk of Magnesia* for medicinal usage.

Although the name may at some point have been owned by GlaxoSmithKline, USPTO registrations show "Milk of Magnesia" and "Phillips' Milk of Magnesia" have both been assigned to Bayer since 1995. In the UK, the non-brand (generic) name of "Milk of Magnesia" and "Phillips' Milk of Magnesia" is "Cream of Magnesia" (Magnesium Hydroxide Mixture, BP).

## As Food Additive

It is added directly to human food, and is affirmed as generally recognized as safe by the FDA. It is known as E number E528.

Magnesium hydroxide is marketed for medical use as chewable tablets, as capsules, powder, and as liquid suspensions, sometimes flavored. These products are sold as antacids to neutralize stomach acid and relieve indigestion and heartburn. It also is a laxative to alleviate constipation. As a laxative, the osmotic force of the magnesia acts to draw fluids from the body. High doses can lead to diarrhea, and can deplete the body's supply of potassium, sometimes leading to muscle cramps.

Some magnesium hydroxide products sold for antacid use (such as Maalox) are formulated to minimize unwanted laxative effects through the inclusion of aluminum hydroxide, which inhibits the contractions of smooth muscle cells in the gastrointestinal tract, thereby counterbalancing the contractions induced by the osmotic effects of the magnesium hydroxide.

## Other Niche uses

Magnesium hydroxide is also a component of antiperspirant. Magnesium hydroxide is useful against canker sores (aphthous ulcer) when used topically.

## Waste Water Treatment

Magnesium hydroxide powder is used industrially to neutralize acidic wastewaters. It also takes part in the Biorock method of building artificial reefs.

## Fire Retardant

Natural magnesium hydroxide (brucite) is used commercially as a fire retardant. Most industrially used magnesium hydroxide is produced synthetically. Like aluminium hydroxide, solid magnesium hydroxide has smoke suppressing and flame retardant properties. This property is attributable to the endothermic decomposition it undergoes at 332 °C (630 °F):

$$Mg(OH)_2 \rightarrow MgO + H_2O$$

The heat absorbed by the reaction retards the fire by delaying ignition of the associated substance. The water released dilutes combustible gases. Common uses of magnesium hydroxide as a flame retardant include additives to cable insulation (i.e. cables for high quality cars, submarines, the Airbus A380, Bugatti Veyron and the PlayStation 4, PlayStation 2, etc.), insulation plastics, roofing (e.g. London Olympic Stadium), and various flame retardant coatings. Other mineral mixtures that are used in similar fire retardant applications are natural mixtures of huntite and hydromagnesite.

## Mineralogy

Brucite crystals (mineral form of $Mg(OH)_2$).

Brucite, the mineral form of $Mg(OH)_2$ commonly found in nature also occurs in the 1:2:1 clay minerals amongst others, in chlorite, in which it occupies the interlayer position normally filled by monovalent and divalent cations such as $Na^+$, $K^+$, $Mg^{2+}$ and $Ca^{2+}$. As a consequence, chlorite interlayers are cemented by brucite and cannot swell nor shrink.

Brucite, in which some of the $Mg^{2+}$ cations have been substituted by $Al^{3+}$ cations, becomes positively charged and constitutes the main basis of layered double hydroxide (LDH). LDH minerals as hydrotalcite are powerful anion sorbents but are relatively rare in nature.

Brucite may also crystallise in cement and concrete in contact with seawater. Indeed, the $Mg^{2+}$ cation is the second most abundant cation in seawater, just behind $Na^+$ and before $Ca^{2+}$. Because brucite is a swelling mineral, it causes a local volumetric expansion responsible for tensile stress in concrete. This leads to the formation of cracks and fissures in concrete, accelerating its degradation in seawater.

For the same reason, dolomite cannot be used as construction aggregate for making concrete. The reaction of magnesium carbonate with the free alkali hydroxides present in the cement porewater also leads to the formation of expansive brucite.

$$MgCO_3 + 2\ NaOH \rightarrow Mg(OH)_2 + Na_2CO_3$$

This reaction, one of the two main alkali-aggregate reaction (AAR) is also known as alkali-carbonate reaction.

# Sodium Fluoride

Sodium fluoride (NaF) is an inorganic compound with the formula NaF. It is a colorless or white solid that is readily soluble in water. It is a common source of fluoride in the production of pharmaceuticals and is used to prevent dental cavities.

In 2016 it was the 215th most prescribed medication in the United States, with more than 2 million prescriptions.

## Uses

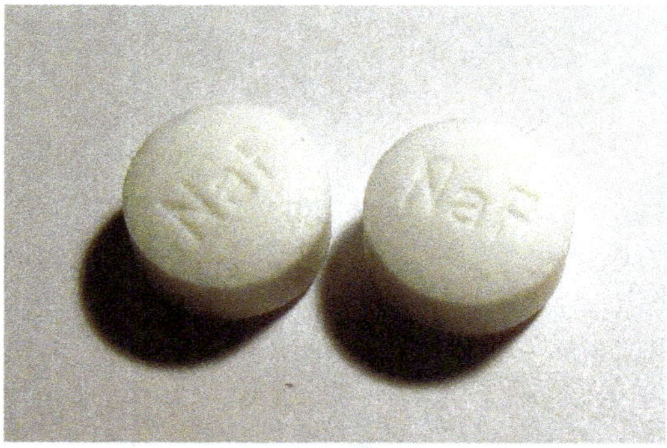


Sodium fluoride is sold in tablets for cavity prevention.

### Dental Caries

Fluoride salts are often added to municipal drinking water (as well as to certain food products in some countries) for the purpose of maintaining dental health. The fluoride enhances the strength of teeth by the formation of fluorapatite, a naturally occurring component of tooth enamel. Although sodium fluoride is used to fluoridate water and, indeed, is the standard by which other water-fluoridation compounds are gauged, hexafluorosilicic acid ($H_2SiF_6$) and its salt sodium hexafluorosilicate ($Na_2SiF_6$) are more commonly used additives in the U.S.

## Osteoporosis

Fluoride supplementation has been extensively studied for the treatment of postmenopausal osteoporosis. This supplementation does not appear to be effective; even though sodium fluoride increases bone density, it does not decrease the risk of fractures.

## Medical Imaging

In medical imaging, fluorine-18-labelled sodium fluoride (USP, sodium fluoride F18) is one of the oldest tracers used in positron emission tomography (PET), having been in use since the 1960s. Relative to conventional bone scintigraphy carried out with gamma cameras or SPECT systems, PET offers more sensitivity and spatial resolution. Fluorine-18 has a half-life of 110 min, which requires it to be used promptly once produced; this logistical limitation hampered its adoption in the face of the more convenient technetium-99m-labelled radiopharmaceuticals. However fluorine-18 is generally considered to be a superior radiopharmaceutical for skeletal imaging. In particular it has a high and rapid bone uptake accompanied by very rapid blood clearance, which results in a high bone-to-background ratio in a short time. Additionally the annihilation photons produced by decay of $^{18}F$ have a high energy of 511-keV compared to 140-keV photons of $^{99m}Tc$.

## Chemistry

Sodium fluoride has a variety of specialty chemical applications in synthesis and extractive metallurgy. It reacts with electrophilic chlorides including acyl chlorides, sulfur chlorides, and phosphorus chloride. Like other fluorides, sodium fluoride finds use in desilylation in organic synthesis. Sodium fluoride can be used to produce fluorocarbons via the Finkelstein reaction; this process has the advantage of being simple to perform on a small scale but is rarely used on an industrial scale due the existence of more effective techniques (e.g. Electrofluorination, Fowler process).

## Other uses

Sodium fluoride is used as a cleaning agent (e.g., as a “laundry sour”). Sodium fluoride is used as a stomach poison for plant-feeding insects. Inorganic fluorides such as fluorosilicates and sodium fluoride complex magnesium ions as magnesium fluorophosphate. They inhibit enzymes such as enolase that require $Mg^{2+}$ as a prosthetic group. Thus, fluoride poisoning prevents phosphate transfer in oxidative metabolism.

## Chemical Structure

Sodium fluoride is an inorganic ionic compound, dissolving in water to give separated $Na^+$ and $F^-$ ions. Like sodium chloride, it crystallizes in a cubic motif where both $Na^+$ and $F^-$ occupy octahedral coordination sites; its lattice spacing, approximately 462 pm, is somewhat smaller than that of sodium chloride.

## Occurrence

The mineral form of NaF, villiaumite, is moderately rare. It is known from plutonic nepheline syenite rocks.

## Production

NaF is prepared by neutralizing hydrofluoric acid or hexafluorosilicic acid ($H_2SiF_6$),both byproducts of the reaction of fluorapatite ($Ca_5(PO_4)_3F$) from phosphate rock during the production of superphosphate fertilizer. Neutralizing agents include sodium hydroxide and sodium carbonate. Alcohols are sometimes used to precipitate the NaF:

$$HF + NaOH \rightarrow NaF + H_2O$$

From solutions containing HF, sodium fluoride precipitates as the bifluoride salt sodium bifluoride ($NaHF_2$). Heating the latter releases HF and gives NaF.

$$HF + NaF \rightleftharpoons NaHF_2$$

In a 1986 report, the annual worldwide consumption of NaF was estimated to be several million tonnes.

# Tetrasulfur Tetranitride

Tetrasulfur tetranitride is an inorganic compound with the formula $S_4N_4$. This gold-poppy coloured solid is the most important binary sulfur nitride, which are compounds that contain only the elements sulfur and nitrogen. It is a precursor to many S-N compounds and has attracted wide interest for its unusual structure and bonding.

Nitrogen and sulfur have similar electronegativities. When the properties of atoms are so highly similar, they often form extensive families of covalently bonded structures and compounds. Indeed, a large number of S-N and S-NH compounds are known with $S_4N_4$ as their parent.

## Structure

$S_4N_4$ adopts an unusual "extreme cradle" structure, with $D_{2d}$ point group symmetry. It can be viewed as a derivative of a hypothetical eight-membered ring of alternating sulfur and nitrogen atoms. The pairs of sulfur atoms across the ring are separated by 2.586 Å, resulting in a cage-like structure as determined by single crystal X-Ray diffraction. The nature of the "transannular" S–S interactions remains a matter of investigation because it is significantly shorter than the sum of the van der Waal's distances but has been explained in the context of molecular orbital theory. The bonding in $S_4N_4$ is considered to be delocalized, which is indicated by the fact that the bond distances between neighboring sulfur and nitrogen atoms are nearly identical. $S_4N_4$ has been shown to co-crystallize with benzene and the $C_{60}$ molecule.

## Properties

$S_4N_4$ is stable to air. It is, however, unstable in the thermodynamic sense with a positive enthalpy of formation of +460 kJ mol$^{-1}$. This endothermic enthalpy of formation originates in the difference in energy of $S_4N_4$ compared to its highly stable decomposition products:

$$2\ S_4N_4 \rightarrow 4\ N_2 + S_8$$

Because one of its decomposition products is a gas, $S_4N_4$ can be used as an explosive. Purer samples tend to be more explosive. Small samples can be detonated by striking with a hammer. $S_4N_4$ is thermochromic, changing from pale yellow below −30 °C to orange at room temperature to deep red above 100 °C.

## Synthesis

S4$N_4$ was first prepared in 1835 by M. Gregory by the reaction of disulfur dichloride with ammonia, a process that has been optimized:

$$6\ S_2Cl_2 + 16\ NH_3 \rightarrow S_4N_4 + S_8 + 12\ NH_4Cl$$

Coproducts of this reaction include heptasulfur imide ($S_7NH$) and elemental sulfur. A related synthesis employs sulfur monochloride and $NH_4Cl$ instead:

$$4\ NH_4Cl + 6\ S_2Cl_2 \rightarrow S_4N_4 + 16\ HCl + S_8$$

An alternative synthesis entails the use of $[(Me_3Si)_2N]_2S$ as a precursor with pre-formed S–N bonds. $[(Me_3Si)_2N]_2S$ is prepared by the reaction of lithium bis(trimethylsilyl)amide and $SCl_2$.

$$2\ [(CH_3)_3Si]_2NLi + SCl_2 \rightarrow [((CH_3)_3Si)_2N]_2S + 2\ LiCl$$

The $[((CH_3)_3Si)_2N]_2S$ reacts with the combination of $SCl_2$ and $SO_2Cl_2$ to form $S_4N_4$, trimethylsilyl chloride, and sulfur dioxide:

$$2[((CH_3)_3Si)_2N]_2S + 2SCl_2 + 2SO_2Cl_2 \rightarrow S_4N_4 + 8\ (CH_3)_3SiCl + 2SO_2$$

## Acid-base Reactions

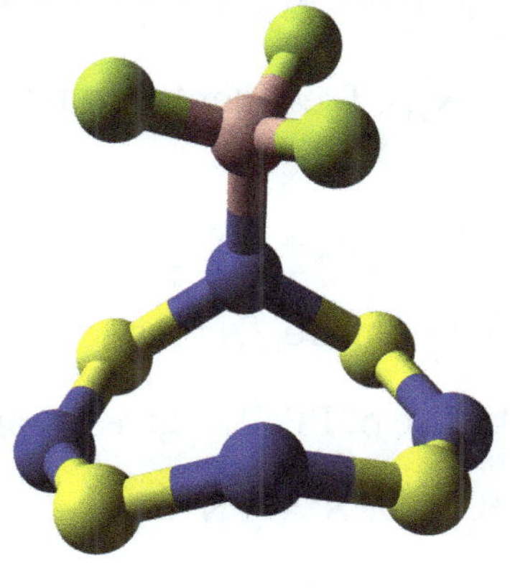

$S_4N_4 \cdot BF_3$

$S_4N_4$ serves as a Lewis base by binding through nitrogen to strongly Lewis acidic compounds such as $SbCl_5$ and $SO_3$. The cage is distorted in these adducts.

$$S_4N_4 + SbCl_5 \rightarrow S_4N_4{\cdot}SbCl_5$$

$$S_4N_4 + SO_3 \rightarrow S_4N_4{\cdot}SO_3$$

The reaction of $[Pt_2Cl_4(PMe_2Ph)_2]$ with $S_4N_4$ is reported to form a complex where a sulfur forms a dative bond to the metal. This compound upon standing is isomerised to a complex in which a nitrogen atom forms the additional bond to the metal centre.

It is protonated by $HBF_4$ to form a tetrafluoroborate salt:

$$S_4N_4 + HBF_4 \rightarrow [S_4N_4H^+][BF^-{}_4]$$

The soft Lewis acid CuCl forms a coordination polymer:

$$n\ S_4N_4 + n\ CuCl \rightarrow (S_4N_4)_n\text{-}\mu\text{-}(\text{-}Cu\text{-}Cl\text{-})_n$$

Dilute NaOH hydrolyzes $S_4N_4$ as follows, yielding thiosulfate and trithionate:

$$2\ S_4N_4 + 6\ OH^- + 9\ H_2O \rightarrow S_2O^{2-}{}_3 + 2\ S_3O^{2-}{}_6 + 8\ NH_3$$

More concentrated base yields sulfite:

$$S_4N_4 + 6\ OH^- + 3\ H_2O \rightarrow S_2O^{2-}{}_3 + 2\ SO^{2-}{}_3 + 4\ NH_3$$

### Metal Complexes

$S_4N_4$ reacts with metal complexes. The cage remains intact in some cases but in other cases, it is degraded. $S_4N_4$ reacts with Vaska's complex ($[Ir(Cl)(CO)(PPh_3)_2]$ in an oxidative addition reaction to form a six coordinate iridium complex where the $S_4N_4$ binds through two sulfur atoms and one nitrogen atom.

## S4N4 as a Precursor to other S-N Compounds

Many S-N compounds are prepared from $S_4N_4$. Reaction with piperidine generates $[S_4N_5]^-$:

$$3\ S_4N_4 + 4\ C_5H_{10}NH \rightarrow (C_5H_{10}NH_2)^+[S_4N_5]^- + (C_5H_{10}N)_2S + 3/8\ S_8 + N_2$$

A related cation is also known, i.e. $[S_4N_5]^+$. Treatment with tetramethylammonium azide produces the heterocycle $[S_3N_3]^-$:

$$S_4N_4 + NMe_4N_3 \rightarrow NMe_4[S_3N_3] + 1/8\ S_8 + 2\ N_2$$

Cyclo-$[S_3N_3]^-$ has 10 pi-electrons: $2e^-/S$ plus $1e^-/N$ plus $1e^-$ for the negative charge.

In an apparently related reaction, the use of $PPN^+N_3$ gives a salt containing the blue $[NS_4]^-$ anion:

$$2\ S_4N_4 + PPN(N_3) \rightarrow PPN[NS_4] + 1/2\ S_8 + 5\ N_2$$

The anion $NS_4^-$ has a chain structure described using the resonance $[S{=}S{=}N{-}S{-}S]^- \leftrightarrow [S{-}S{-}N{=}S{=}S]^-$.

$S_4N_4$ reacts with electron-poor alkynes.

Chlorination of $S_4N_4$ gives thiazyl chloride.

Passing gaseous $S_4N_4$ over silver metal yields the low temperature superconductor polythiazyl or polysulfurnitride (transition temperature (0.26±0.03) K), often simply called "$(SN)_x$". In the conversion, the silver first becomes sulfided, and the resulting $Ag_2S$ catalyzes the conversion of the $S_4N_4$ into the four-membered ring $S_2N_2$, which readily polymerizes.

$$S_4N_4 + 8\,Ag \rightarrow 4\,Ag_2S + 2\,N_2$$

$$S_4N_4 \rightarrow (SN)_x$$

### $Se_4N_4$

The selenium compound $Se_4N_4$ is known and has been the subject of some research. In addition, adducts of aluminium chloride with $Se_2N_2$ have been isolated; this is formed from $Se_4N_4$.

## Aluminium Fluoride

Aluminium fluoride refers to inorganic compounds with the formula $AlF_3 \cdot xH_2O$. They are all colorless solids. Anhydrous $AlF_3$ is used in the production of aluminium metal.

### Occurrence and Production

Aside from anhydrous $AlF_3$, several hydrates are known. With the formula $AlF_3 \cdot xH_2O$, these compounds include monohydrate (x = 1), two polymorphs of the trihydrate (x = 3), a hexahydrate (x = 6), and a nonahydrate (x = 9).

The majority of aluminium fluoride is produced by treating alumina with hydrogen fluoride at 700 °C: Fluorosilicic acid may also be used make aluminum fluoride.

$$H_2SiF_6 + Al_2O_3 + 3\,H_2O \rightarrow 2\,AlF_3 + SiO_2 + 4\,H_2O$$

Alternatively, it is manufactured by thermal decomposition of ammonium hexafluoroaluminate. For small scale laboratory preparations, $AlF_3$ can also be prepared by treating aluminium hydroxide or aluminium metal with hydrogen fluoride.

Aluminium fluoride trihydrate is found in nature as the rare mineral rosenbergite. The non-hydrated form appears as the mineral oskarssonite.

### Structure

According to X-ray crystallography, anhydrous $AlF_3$ adopts the rhenium trioxide motif, featuring distorted $AlF_6$ octahedra. Each fluoride is connected to two Al centers. Because of its 3-dimensional polymeric structure, $AlF_3$ has a high melting point. The other trihalides of aluminium in the solid state differ, $AlCl_3$ has a layer structure and $AlBr_3$ and $AlI_3$, are molecular dimers. Also they have low melting points and evaporate readily to give dimers. In the gas phase aluminium fluoride exists as trigonal molecules of $D_{3h}$ symmetry. The Al-F bond lengths of this gaseous molecule are 163 pm.

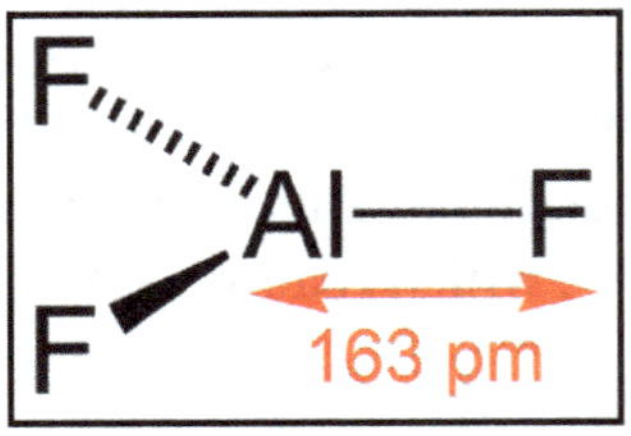


Like most gaseous metal trifluorides, $AlF_3$ adopts a planar structure upon evaporation.

## Applications

Aluminium fluoride is an important additive for the production of aluminium by electrolysis. Together with cryolite, it lowers the melting point to below 1000 °C and increases the conductivity of the solution. It is into this molten salt that aluminium oxide is dissolved and then electrolyzed to give bulk Al metal.

Aluminum fluoride complexes are used to study the mechanistic aspects of phosphoryl transfer reactions in biology, which are of fundamental importance to cells, as phosphoric acid anhydrides such as ATP and GTP control most of the reactions involved in metabolism, growth and differentiation. The observation that aluminum fluoride can bind to and activate heterotrimeric G proteins has proven to be useful for the study of G protein activation in vivo, for the elucidation of three-dimensional structures of several GTPases, and for understanding the biochemical mechanism of GTP hydrolysis, including the role of GTPase-activating proteins.

### Niche uses

Together with zirconium fluoride, aluminium fluoride is an ingredient for the production of fluoroaluminate glasses.

It is also used to inhibit fermentation.

Like magnesium fluoride it used as a low-index optical thin film, particularly when far UV transparency is required. Its deposition by physical vapor deposition, particularly by evaporation, is favorable.

# Barium Sulfate

Barium sulfate (or sulphate) is the inorganic compound with the chemical formula $BaSO_4$. It is a white crystalline solid that is odorless and insoluble in water. It occurs as the mineral barite, which is the main commercial source of barium and materials prepared from it. The white opaque appearance and its high density are exploited in its main applications.

## Uses

### Drilling Fluids

About 80% of the world's barium sulfate production, mostly purified mineral, is consumed as a component of oil well drilling fluid. It increases the density of the fluid, increasing the hydrostatic pressure in the well and reducing the chance of a blowout.

## Radiocontrast Agent

Barium sulfate in suspension is frequently used medically as a radiocontrast agent for X-ray imaging and other diagnostic procedures. It is most often used in imaging of the GI tract during what is colloquially known as a "barium meal". It is administered orally, or by enema, as a suspension of fine particles in a thick milk like solution (often with sweetening and flavoring agents added). Although barium is a heavy metal, and its water-soluble compounds are often highly toxic, the low solubility of barium sulfate protects the patient from absorbing harmful amounts of the metal. Barium sulfate is also readily removed from the body, unlike Thorotrast, which it replaced. Due to the relatively high atomic number ($Z = 56$) of barium, its compounds absorb X-rays more strongly than compounds derived from lighter nuclei.

## Pigment

The majority of synthetic barium sulfate is used as a component of white pigment for paints. In oil paint, barium sulfate is almost transparent, and is used as a filler or to modify consistency. One major manufacturer of artists' oil paint sells "permanent white" that contains a mixture of titanium white pigment ($TiO_2$) and barium sulfate. The combination of barium sulfate and zinc sulfide ($ZnS$) is the inorganic pigment called lithopone. In photography it is used as a coating for certain photographic papers.

## Paper Brightener

A thin layer of barium sulfate called baryta is first coated on the base surface of most photographic paper to increase the reflectiveness of the image, with the first such paper introduced in 1884 in Germany. The light-sensitive silver halide emulsion is then coated over the baryta layer. The baryta coating limits the penetration of the emulsion into the fibers of the paper and makes the emulsion more even, resulting in more uniform blacks. More recently, baryta has been used to brighten papers intended for ink-jet printing.

## Plastics Filler

Barium sulfate is commonly used as a filler for plastics to increase the density of the polymer in vibrational mass damping applications. In polypropylene and polystyrene plastics, it is used as a filler in proportions up to 70%. It has an effect of increasing acid and alkali resistance and opacity.

## Niche uses

Barium sulfate is used in soil testing. Tests for soil pH and other qualities of soil use colored indicators, and small particles (usually clay) from the soil can cloud the test mixture and make it hard to see the color of the indicator. Barium sulfate added to the mixture binds with these particles, making them heavier so they fall to the bottom, leaving a clearer solution.

In colorimetry, barium sulfate is used as a near-perfect diffuser when measuring light sources.

In metal casting, the moulds used are often coated with barium sulfate in order to prevent the molten metal from bonding with the mould.

It is also used in brake linings, anacoustic foams, powder coatings, and root canal filling.

### Catalyst Support

Barium sulfate is used as a catalyst support when selectively hydrogenating functional groups that are sensitive to overreduction. With a low surface area, the contact time of the substrate with the catalyst is shorter and thus selectivity is achieved. Palladium on barium sulfate is also used as a catalyst in the Rosenmund reduction.

### Pyrotechnics

As barium compounds emit a green light when burned, barium salts are often used in green pyrotechnic formulas, although nitrate and chlorate salts are more common. Barium sulfate is commonly used as a component of "strobe" pyrotechnic compositions.

### Copper Industry

As barium sulfate has high burning point and is insoluble in water, it is used as a coating material in casting of copper anode plates. The anode plates are cast in copper molds, so to avoid the contact of the liquid copper and the solid copper mold, a solution of barium sulfate in water is used as a coating material on the mold surface. Thus when the liquid copper solidifies in form of an anode plate it can be easily released from its mold.

### Radiometric Measurements

Barium sulfate is sometimes used (or else PTFE) to coat the interior of integrating spheres due to the high reflectance of the material and near Lambertian characteristics.

### Production

Almost all of the barium consumed commercially is obtained from barite, which is often highly impure. Barite is processed by carbothermal reduction (heating with coke) to give barium sulfide:

$$BaSO_4 + 4\,C \rightarrow BaS + 4\,CO$$

In contrast to barium sulfate, barium sulfide is soluble in water and readily converted to the oxide, carbonate, and halides. To produce highly pure barium sulfate, the sulfide or chloride is treated with sulfuric acid or sulfate salts:

$$BaS + H_2SO_4 \rightarrow BaSO_4 + H_2S$$

Barium sulfate produced in this way is often called *blanc fixe*, which is French for "permanent white." Blanc fixe is the form of barium encountered in consumer products, such as paints.

In the laboratory barium sulfate is generated by combining solutions of barium ions and sulfate salts. Because barium sulfate is the least toxic salt of barium due to its insolubility, wastes containing barium salts are sometimes treated with sodium sulfate to immobilize (detoxify) the barium. Barium sulfate is one of the most insoluble salts of sulfate. Its low solubility is exploited in qualitative inorganic analysis as a test for $Ba^{2+}$ ions as well as for sulfate.

# Beryllium Oxide

Beryllium oxide (BeO), also known as beryllia, is an inorganic compound with the formula BeO. This colourless solid is a notable electrical insulator with a higher thermal conductivity than any other non-metal except diamond, and exceeds that of most metals. As an amorphous solid, beryllium oxide is white. Its high melting point leads to its use as a refractory material. It occurs in nature as the mineral bromellite. Historically and in materials science, beryllium oxide was called glucina or glucinium oxide.

## Preparation and Chemical Properties

Beryllium oxide can be prepared by calcining (roasting) beryllium carbonate, dehydrating beryllium hydroxide, or igniting metallic beryllium:

$$BeCO_3 \rightarrow BeO + CO_2$$

$$Be(OH)_2 \rightarrow BeO + H_2O$$

$$2\ Be + O_2 \rightarrow 2\ BeO$$

Igniting beryllium in air gives a mixture of BeO and the nitride $Be_3N_2$. Unlike the oxides formed by the other group 2 elements (alkaline earth metals), beryllium oxide is amphoteric rather than basic.

Beryllium oxide formed at high temperatures (>800 °C) is inert, but dissolves easily in hot aqueous ammonium bifluoride ($NH_4HF_2$) or a solution of hot concentrated sulfuric acid ($H_2SO_4$) and ammonium sulfate ($(NH_4)_2SO_4$).

## Structure

BeO crystallizes in the hexagonal wurtzite structure, featuring tetrahedral $Be^{2+}$ and $O^{2-}$ centres, like lonsdaleite and w-BN (with both of which it is isoelectronic). In contrast, the oxides of the larger group-2 metals, i.e., MgO, CaO, SrO, BaO, crystallize in the cubic rock salt motif with octahedral geometry about the dications and dianions. At high temperature the structure transforms to a tetragonal form.

In the vapour phase, beryllium oxide is present as discrete diatomic molecules. In the language of valence bond theory, these molecules can be described as adopting *sp* orbital hybridisation on both atoms, featuring one σ (between one *sp* orbital on each atom) and one π bond (between aligned *p* orbitals on each atom oriented perpendicular to the molecular axis). Molecular orbital theory provides a slightly different picture with no *net* sigma bonding (because the 2*s* orbitals of the two atoms combine to form a filled sigma bonding orbital and a filled sigma* anti-bonding orbital) and two pi bonds formed between both pairs of *p* orbitals oriented perpendicular to the molecular axis. The sigma orbital formed by the *p* orbitals aligned along the molecular axis is unfilled. The corresponding ground state is ...$(2s\sigma)^2(2s\sigma^*)^2(2p\pi)^4$ (as in the isoelectronic $C_2$ molecule), where both bonds can be considered as dative bonds from oxygen towards beryllium.

## Applications

High-quality crystals may be grown hydrothermally, or otherwise by the Verneuil method. For the most part, beryllium oxide is produced as a white amorphous powder, sintered into larger shapes. Impurities, like carbon, can give a variety of colours to the otherwise colourless host crystals.

Sintered beryllium oxide is a very stable ceramic. Beryllium oxide is used in rocket engines and as a transparent protective over-coating on aluminised telescope mirrors.

Beryllium oxide is used in many high-performance semiconductor parts for applications such as radio equipment because it has good thermal conductivity while also being a good electrical insulator. It is used as a filler in some thermal interface materials such as thermal grease. Some power semiconductor devices have used beryllium oxide ceramic between the silicon chip and the metal mounting base of the package to achieve a lower value of thermal resistance than a similar construction of aluminium oxide. It is also used as a structural ceramic for high-performance microwave devices, vacuum tubes, magnetrons, and gas lasers. BeO has been proposed as a neutron moderator for naval marine high-temperature gas-cooled reactors (MGCR), as well as NASA's Kilopower nuclear reactor for space applications.

# Beryllium Hydride

Beryllium hydride (systematically named poly[beryllane(2)] and beryllium dihydride) is an inorganic compound with the chemical formula $(BeH_2)_n$ (also written $([BeH_2])_n$ or $BeH_2$). This alkaline earth hydride is a colourless solid that is insoluble in solvents that do not decompose it. Unlike the ionically bonded hydrides of the heavier Group 2 elements, beryllium hydride is covalently bonded (three-center two-electron bond).

## Synthesis

Unlike the other group 2 metals, beryllium does not react with hydrogen. Instead, $BeH_2$ is prepared from preformed beryllium(II) compounds. It was first synthesised in 1951 by treating dimethylberyllium, $Be(CH_3)_2$, with lithium aluminium hydride, $LiAlH_4$.

Purer $BeH_2$ forms from the pyrolysis of di-tert-butylberyllium, $Be(C(CH_3)_3)_2$ at 210 °C.

A route to highly pure samples involve the reaction of triphenylphosphine, $PPh_3$, with beryllium borohydride, $Be(BH_4)_2$:

$$Be(BH_4)_2 + 2\ PPh_3 \rightarrow 2\ Ph_3PBH_3 + BeH_2$$

## Structure

$BeH_2$ is usually formed as an amorphous white solid, but a hexagonal crystalline form with a higher density (~0.78 g $cm^{-3}$) was reported, prepared by heating amorphous $BeH_2$ under pressure, with 0.5-2.5% LiH as a catalyst.

A more recent investigation found that crystalline beryllium hydride has a body-centred orthorhombic unit cell, containing a network of corner-sharing $BeH_4$ tetrahedra, in contrast to the flat, hydrogen-bridged, infinite chains previously thought to exist in crystalline $BeH_2$.

Studies of the amorphous form also find that it consists of a network of corner shared tetrahedra.

## Chemical Properties

### Reaction with Water and Acids

Beryllium hydride reacts slowly with water but is rapidly hydrolysed by acid such as hydrogen chloride to form beryllium chloride.

### Reaction with Lewis Bases

Beryllium hydride reacts with trimethylamine, $N(CH_3)_3$ to form a dimeric adduct, with bridging hydrides. However, with dimethylamine, $HN(CH_3)_2$ it forms a trimeric beryllium diamide, $[Be(N(CH_3)_2)_2]_3$ and hydrogen. The reaction with lithium hydride where the hydride ion is the Lewis base, forms sequentially $LiBeH_3$ and $Li_2BeH_4$.

## Dihydridoberyllium

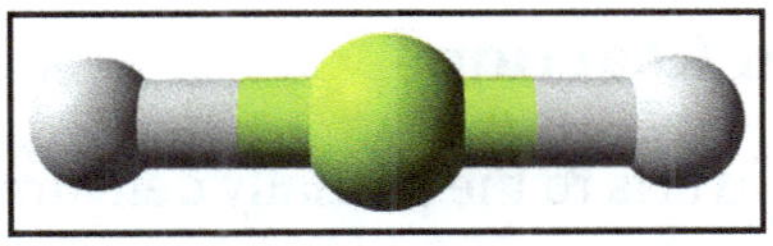

Structure of gaseous $BeH_2$.

Dihydridoberyllium is a related compound with the chemical formula $BeH_2$ (also written $[BeH_2]$). It is a gas that cannot persist undiluted. Unsolvated dihydridoberyllium will spontaneously autopolymerise to oligomers. Free molecular $BeH_2$ produced by electrical discharge at high temperature has been confirmed as linear with a Be-H bond length of 133.376 pm. Its hybridisation is sp.

## Chemical properties

In theory, the two-coordinate hydridoberyllium group (-BeH) in hydridoberylliums such as dihydridoberyllium can accept an electron-pair donating ligand into the molecule by adduction:

$$[BeH_2] + L \rightarrow [BeH_2L]$$

Because of this acceptance of the electron-pair donating ligand (L), dihydridoberyllium has Lewis-acidic character. Dihydridoberyllium can accept two electron pairs from ligands, as in the case of the tetrahydridoberyllate(2-) anion ($BeH^{2-}_4$).

# Calcium Sulfate

Calcium sulfate (or calcium sulphate) is the inorganic compound with the formula $CaSO_4$ and related hydrates. In the form of γ-anhydrite (the anhydrous form), it is used as a desiccant. One particular hydrate is better known as plaster of Paris, and another occurs naturally as the mineral gypsum. It has many uses in industry. All forms are white solids that are poorly soluble in water. Calcium sulfate causes permanent hardness in water.

## Hydration States and Crystallographic Structures

The compound exists in three levels of hydration corresponding to different crystallographic structures and to different minerals in nature:

- $CaSO_4$ (anhydrite): Anhydrous state.
- $CaSO_4 \cdot 2\ H_2O$ (gypsum and selenite (mineral)): Dihydrate.
- $CaSO_4 \cdot \frac{1}{2}\ H_2O$ (bassanite): Hemihydrate, also known as plaster of Paris. Specific hemihydrates are sometimes distinguished: α-hemihydrate and β-hemihydrate.

### Uses

The main use of calcium sulfate is to produce plaster of Paris and stucco. These applications exploit the fact that calcium sulfate which has been powdered and calcined forms a moldable paste upon hydration and hardens as crystalline calcium sulfate dihydrate. It is also convenient that calcium sulfate is poorly soluble in water and does not readily dissolve in contact with water after its solidification.

## Hydration and Dehydration Reactions

With judicious heating, gypsum converts to the partially dehydrated mineral called calcium sulfate hemihydrate, calcined gypsum, and plaster of Paris.This material has the formula $CaSO_4{\cdot}(nH_2O)$, where $0.5 \leq n \leq 0.8$. Temperatures between 100 and 150 °C (212–302 °F) are required to drive off the water within its structure. The details of the temperature and time depend on ambient humidity. Temperatures as high as 170 °C (338 °F) are used in industrial calcination, but at these temperatures γ-anhydrite begins to form. The heat energy delivered to the gypsum at this time (the heat of hydration) tends to go into driving off water (as water vapor) rather than increasing the temperature of the mineral, which rises slowly until the water is gone, then increases more rapidly. The equation for the partial dehydration is:

$$CaSO_4 \cdot 2\ H_2O \ \rightarrow \ CaSO_4 \cdot \frac{1}{2} H_2O + 1\frac{1}{2} H_2O \uparrow$$

The endothermic property of this reaction is relevant to the performance of drywall, conferring fire resistance to residential and other structures. In a fire, the structure behind a sheet of drywall will remain relatively cool as water is lost from the gypsum, thus preventing (or substantially retarding) damage to the framing (through combustion of wood members or loss of strength of steel at high temperatures) and consequent structural collapse. But at higher temperatures, calcium sulfate will release oxygen and act as an oxidizing agent. This property is used in aluminothermy. In contrast to most minerals, which when rehydrated simply form liquid or semi-liquid pastes, or remain powdery, calcined gypsum has an unusual property: When mixed with water at normal (ambient) temperatures, it quickly reverts chemically to the preferred dihydrate form, while physically "setting" to form a rigid and relatively strong gypsum crystal lattice:

$$CaSO_4 \cdot \frac{1}{2}H_2O + 1\frac{1}{2}H_2O \rightarrow CaSO_4 \cdot 2\,H_2O$$

This reaction is exothermic and is responsible for the ease with which gypsum can be cast into various shapes including sheets (for drywall), sticks (for blackboard chalk), and molds (to immobilize broken bones, or for metal casting). Mixed with polymers, it has been used as a bone repair cement. Small amounts of calcined gypsum are added to earth to create strong structures directly from cast earth, an alternative to adobe (which loses its strength when wet). The conditions of dehydration can be changed to adjust the porosity of the hemihydrate, resulting in the so-called α- and β-hemihydrates (which are more or less chemically identical).

On heating to 180 °C (356 °F), the nearly water-free form, called γ-anhydrite ($CaSO_4 \cdot nH_2O$ where $n$ = 0 to 0.05) is produced. γ-Anhydrite reacts slowly with water to return to the dihydrate state, a property exploited in some commercial desiccants. On heating above 250 °C, the completely anhydrous form called β-anhydrite or "natural" anhydrite is formed. Natural anhydrite does not react with water, even over geological timescales, unless very finely ground.

The variable composition of the hemihydrate and γ-anhydrite, and their easy inter-conversion, is due to their nearly identical crystal structures containing "channels" that can accommodate variable amounts of water, or other small molecules such as methanol.

## Food Industry

The calcium sulfate hydrates are used as a coagulant in products such as tofu.

For the FDA, it is permitted in Cheese and Related Cheese Products; Cereal Flours; Bakery Products; Frozen Desserts; Artificial Sweeteners for Jelly & Preserves; Condiment Vegetables; and Condiment Tomatoes and some candies.

It is known in the E number series as E516, and the FAO knows it as a firming agent, a flour treatment agent, a sequestrant, and a leavening agent.

## Other uses

When sold at the anhydrous state as a desiccant with a color-indicating agent under the name Drierite, it appears blue (anhydrous) or pink (hydrated) due to impregnation with cobalt(II) chloride, which functions as a moisture indicator.

Up to the 1970s, commercial quantities of sulfuric acid were produced in Whitehaven (Cumbria, UK) from anhydrous calcium sulfate. Upon being mixed with shale or marl, and roasted, the sulfate liberates sulfur trioxide gas, a precursor in sulfuric acid production, the reaction also produces calcium silicate, a mineral phase essential in cement clinker production.

$$CaSO_4 + SiO_2 \rightarrow CaSiO_3 + SO_3$$

Drierite

## Production and Occurrence

The main sources of calcium sulfate are naturally occurring gypsum and anhydrite, which occur at many locations worldwide as evaporites. These may be extracted by open-cast quarrying or by deep mining. World production of natural gypsum is around 127 million tonnes per annum.

In addition to natural sources, calcium sulfate is produced as a by-product in a number of processes:

- In flue-gas desulfurization, exhaust gases from fossil-fuel power stations and other processes (e.g. cement manufacture) are scrubbed to reduce their sulfur oxide content, by injecting finely ground limestone or lime. This produces an impure calcium sulfite, which oxidizes on storage to calcium sulfate.
- In the production of phosphoric acid from phosphate rock, calcium phosphate is treated with sulfuric acid and calcium sulfate precipitates.
- In the production of hydrogen fluoride, calcium fluoride is treated with sulfuric acid, precipitating calcium sulfate.
- In the refining of zinc, solutions of zinc sulfate are treated with lime to co-precipitate heavy metals such as barium.
- Calcium sulfate can also be recovered and re-used from scrap drywall at construction sites.

These precipitation processes tend to concentrate radioactive elements in the calcium sulfate product. This issue is particular with the phosphate by-product, since phosphate ores naturally contain uranium and its decay products such as radium-226, lead-210 and polonium-210.

Calcium sulfate is also a common component of fouling deposits in industrial heat exchangers, because its solubility decreases with increasing temperature.

## Retrograde Solubility

The dissolution of the different crystalline phases of calcium sulfate in water is exothermic and release heat (decrease in Enthalpy: $\Delta H < 0$). As an immediate consequence, to proceed, the dissolution reaction needs to evacuate this heat that can be considered as a product of reaction. If the system is cooled, the dissolution equilibrium will evolve towards the right according to the Le Chatelier principle and calcium sulfate will dissolve more easily. The solubility of calcium sulfate increases thus when the temperature decreases. If the temperature of the system is raised, the reaction heat cannot dissipate and the equilibrium will regress towards the left according to Le Chatelier principle. The solubility of calcium sulfate decreases thus when temperature increases. This contra-intuitive solubility behaviour is called retrograde solubility. It is less common than for most of the salts whose dissolution reaction is endothermic (i.e., the reaction consumes heat: increase in Enthalpy: $\Delta H > 0$) and whose solubility increases with temperature. Another calcium compound, calcium hydroxide ($Ca(OH)_2$, portlandite) also exhibits a retrograde solubility for the same thermodynamic reason: because its dissolution reaction is also exothermic and releases heat. So, to dissolve higher amounts of calcium sulfate or calcium hydroxide in water, it is necessary to cool down the solution close to its freezing point instead of increasing its temperature.

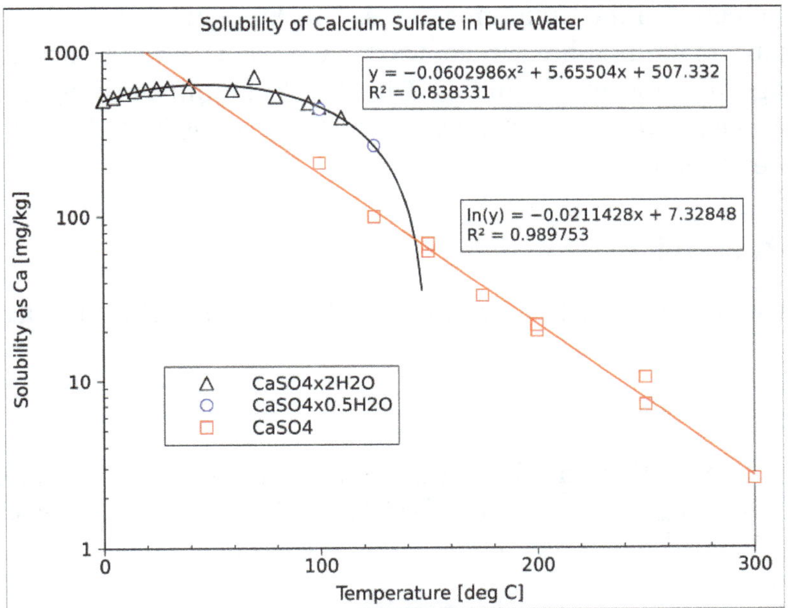


Temperature dependence of the solubility of calcium sulfate (3 phases) in pure water.

The retrograde solubility of calcium sulfate is also responsible for its precipitation in the hottest zone of heating systems and for its contribution to the formation of scale in boilers along with the precipitation of calcium carbonate whose solubility also decreases when $CO_2$ degasses from hot water or can escape out of the system.

## Titanium (III) Chloride

Titanium(III) chloride is the inorganic compound with the formula $TiCl_3$. At least four distinct species have this formula; additionally hydrated derivatives are known. $TiCl_3$ is one of the most common halides of titanium and is an important catalyst for the manufacture of polyolefins.

## Structure and Bonding

In $TiCl_3$, each Ti atom has one *d* electron, rendering its derivatives paramagnetic, i.e. the substance is attracted into a magnetic field. Solutions of titanium(III) chloride are violet, which arises from excitations of its d-electron. The colour is not very intense since the transition is forbidden by the Laporte selection rule.

Four solid forms or polymorphs of $TiCl_3$ are known. All feature titanium in an octahedral coordination sphere. These forms can be distinguished by crystallography as well as by their magnetic properties, which probes exchange interactions. β-$TiCl_3$ crystallizes as brown needles. Its structure consists of chains of $TiCl_6$ octahedra that share opposite faces such that the closest Ti—Ti contact is 2.91 Å. This short distance indicates strong metal-metal interactions. The three violet "layered" forms, named for their color and their tendency to flake, are called alpha, gamma, and delta. In α-$TiCl_3$, the chloride anions are hexagonal close-packed. In γ-$TiCl_3$, the chlorides anions are cubic close-packed. Finally, disorder in shift successions, causes an intermediate between alpha and gamma structures, called the delta (δ) form. The $TiCl_6$ share edges in each form, with 3.60 Å being the shortest distance between the titanium cations. This large distance between titanium cations precludes direct metal-metal bonding. In contrast, the trihalides of the heavier metals hafnium and zirconium engage in metal-metal bonding. Direct Zr-Zr bonding is indicated in zirconium(III) chloride. The difference between the Zr(III) and Ti(III) materials is attributed in part to the relative radii of these metal centers.

## Synthesis and Reactivity

$TiCl_3$ is produced usually by reduction of titanium(IV) chloride. Older reduction methods used hydrogen:

$$2\ TiCl_4 + H_2 \rightarrow 2\ HCl + 2\ TiCl_3$$

It is conveniently reduced with aluminium and sold as a mixture with aluminium trichloride, $TiCl_3 \cdot AlCl_3$. This mixture can be separated to afford $TiCl_3(THF)_3$. The complex adopts a meridional structure.

Its hydrate can be synthesised by dissolving titanium in aqueous hydrochloric acid.

$$2\ Ti + 6\ HCl + 6\ H_2O \rightarrow 2\ TiCl_3(H_2O)_3 + 3\ H_2$$

$TiCl_3$ forms a variety of coordination complexes, most of which are octahedral. The light-blue crystalline adduct $TiCl_3(THF)_3$ forms when $TiCl_3$ is treated with tetrahydrofuran.

$$TiCl_3 + 3\ C_4H_8O \rightarrow TiCl_3(OC_4H_8)_3$$

An analogous dark green complex arises from complexation with dimethylamine. In a reaction where all ligands are exchanged, $TiCl_3$ is a precursor to the tris acetylacetonate complex.

The more reduced titanium(II) chloride is prepared by the thermal disproportionation of $TiCl_3$ at 500 °C. The reaction is driven by the loss of volatile $TiCl_4$:

$$2\ TiCl_3 \rightarrow TiCl_2 + TiCl_4$$

The ternary halides, such as $A_3TiCl_6$, have structures that depend on the cation ($A^+$) added. Caesium chloride treated with titanium(II) chloride and hexachlorobenzene produces crystalline $CsTi_2Cl_7$. In these structures $Ti^{3+}$ exhibits octahedral coordination geometry.

### Applications

$TiCl_3$ is the main Ziegler–Natta catalyst, responsible for most industrial production of polyethylene. The catalytic activities depend strongly on the polymorph of the TiCl3 (α vs. β vs. γ vs. δ) and the method of preparation.

### Laboratory use

$TiCl_3$ is also a specialized reagent in organic synthesis, useful for reductive coupling reactions, often in the presence of added reducing agents such as zinc. It reduces oximes to imines. Titanium trichloride can reduce nitrate to ammonium ion thereby allowing for the sequential analysis of nitrate and ammonia. Slow deterioration occurs in air-exposed titanium trichloride, often resulting in erratic results, e.g. in reductive coupling reactions.

## Sodium Amide

Sodium amide, commonly called sodamide (systematic name sodium azanide), is the inorganic compound with the formula $NaNH_2$. It is a salt composed of the sodium cation and the azanide anion. This solid, which is dangerously reactive toward water, is white, but commercial samples are typically gray due to the presence of small quantities of metallic iron from the manufacturing process. Such impurities do not usually affect the utility of the reagent. $NaNH_2$ conducts electricity in the fused state, its conductance being similar to that of NaOH in a similar state. $NaNH_2$ has been widely employed as a strong base in organic synthesis.

### Preparation and Structure

Sodium amide can be prepared by the reaction of sodium with ammonia gas, but it is usually prepared by the reaction in liquid ammonia using iron(III) nitrate as a catalyst. The reaction is fastest at the boiling point of the ammonia, c. −33 °C. An electride, $[Na(NH_3)_6]^+e^-$, is formed as a reaction intermediate.

$$2\ Na + 2\ NH_3 \rightarrow 2\ NaNH_2 + H_2$$

$NaNH_2$ is a salt-like material and as such, crystallizes as an infinite polymer. The geometry about sodium is tetrahedral. In ammonia, $NaNH_2$ forms conductive solutions, consistent with the presence of $Na(NH_3)_6^+$ and $NH_2^-$ ions.

### Uses

Sodium amide is mainly used as a strong base in organic chemistry, often in liquid ammonia solution. It is the reagent of choice for the drying of ammonia (liquid or gaseous). One of the main advantages to the use of sodium amide is that it mainly functions as a nucleophile. In the industrial production of indigo, sodium amide is a component of the highly basic mixture that induces cyclisation of *N*-phenylglycine. The reaction produces ammonia, which is recycled typically.

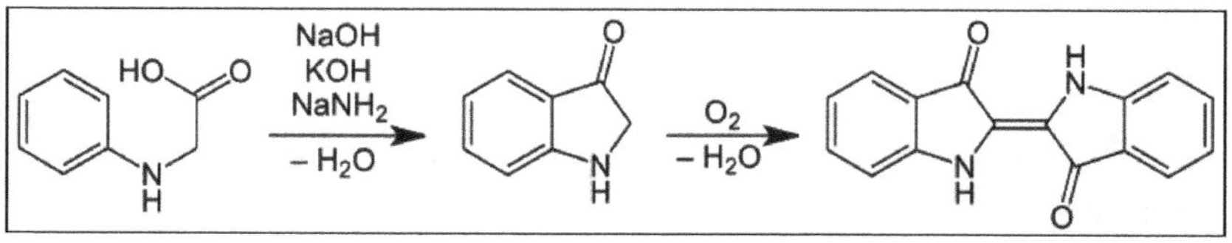


Pfleger's synthesis of indigo dye.

## Dehydrohalogenation

Sodium amide induces the loss of two equivalents of hydrogen bromide from a vicinal dibromoalkane to give a carbon-carbon triple bond, as in a preparation of phenylacetylene. Usually two equivalents of sodium amide yields the desired alkyne. Three equivalents are necessary in the preparation of a terminal alkynes because the terminal CH of the resulting alkyne protonates an equivalent amount of base.

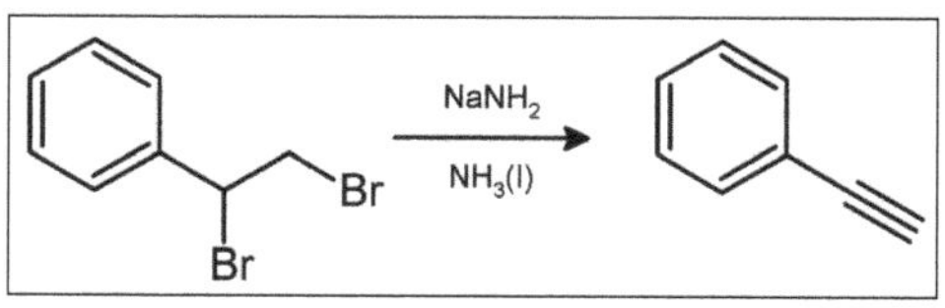


Hydrogen chloride and ethanol can also be eliminated in this way, as in the preparation of 1-ethoxy-1-butyne.

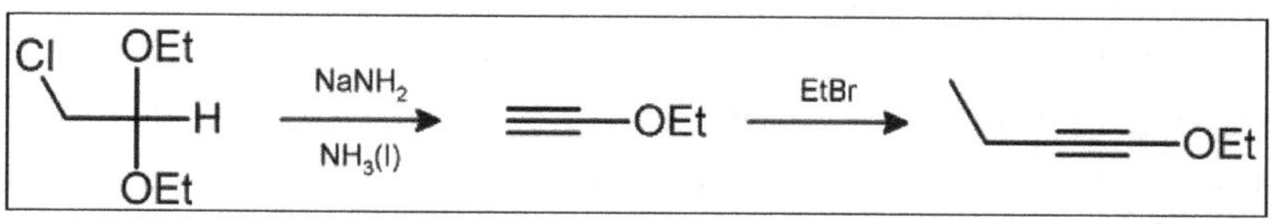


## Cyclization Reactions

Where there is no β-hydrogen to be eliminated, cyclic compounds may be formed, as in the preparation of methylenecyclopropane below.

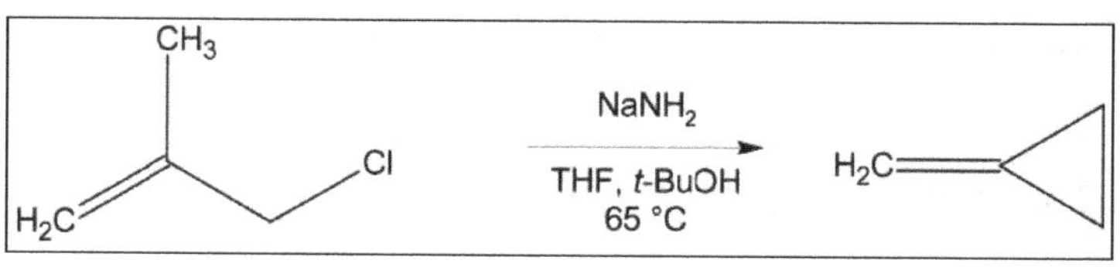


Cyclopropenes, aziridines and cyclobutanes may be formed in a similar manner.

## Deprotonation of Carbon and Nitrogen Acids

Carbon acids which can be deprotonated by sodium amide in liquid ammonia include terminal alkynes, methyl ketones, cyclohexanone, phenylacetic acid and its derivatives and diphenylmethane. Acetylacetone loses two protons to form a dianion. Sodium amide will also deprotonate indole and piperidine.

### Related Non-nucleophilic bases

It is however poorly soluble in solvents other than ammonia. Its use has been superseded by the related reagents sodium hydride, sodium bis(trimethylsilyl)amide (NaHMDS), and lithium diisopropylamide (LDA).

### Other Reactions

- Rearrangement with orthodeprotonation.
- Oxirane synthesis.
- Indole synthesis.
- Chichibabin reaction.

## Inorganic Hydroxides

The hydroxide moiety contains a oxygen-hydrogen bond, carrying a negative charge. Some of the inorganic hydroxides, like sodium hydroxide, calcium hydroxide, and other alkali hydroxides, are strong bases that are important industrial alkalis. Some of them, like hydroxides of zinc, aluminium, tin, beryllium, and lead, are amphoteric, as they act like both acids and bases.

Lithium hydroxide is used in breathing gas purification systems for spacecraft, submarines, and rebreathers to remove carbon dioxide from exhaled gas. Alkaline earth metal compounds, like magnesium hydroxide, calcium hydroxide, strontium hydroxide, and barium hydroxide, are strong bases. A solution of calcium hydroxide and limewater can be used to test for carbon dioxide. Soda lime, which is a mixture of NaOH and Ca(OH)2, is used as a CO2 absorbent. Unlike the alkali and alkaline earth hydroxides, boron hydroxide, B(OH)3, known as boric acid, is an acid that does not dissociate in aqueous solution. Basic aluminium hydroxide is also amphoteric. Gallium hydroxide, indium hydroxide, and thallium (III) hydroxide are also amphoteric.

The hydroxide ion acts as a base catalyst in several organic transformations. Hydroxide can also act as a Lewis-base catalyst, and as a ligand. Organic reactions where hydroxide can act as a nucleophilic reagent are during amide hydrolysis, in the Cannizzaro reaction, during nucleophilic aliphatic substitution, during nucleophilic aromatic substitution, and in elimination reactions. Alkali metal hydroxides, such as sodium hydroxide, are used in the manufacture of pulp and paper, textiles, soaps, and detergents. Potassium hydroxide is used in agriculture, batteries, petroleum and natural gas refining, and manicure treatments.

### Lithium Hydroxide

Lithium hydroxide is an inorganic compound with the formula LiOH. It is a white hygroscopic crystalline material. It is soluble in water and slightly soluble in ethanol, and is available commercially in anhydrous form and as the monohydrate ($LiOH{\cdot}H_2O$), both of which are strong bases. It is the weakest base among the alkali metal hydroxides.

## Production and Reactions

Lithium hydroxide is produced in a metathesis reaction between lithium carbonate and calcium hydroxide:

$$Li_2CO_3 + Ca(OH)_2 \rightarrow 2\ LiOH + CaCO_3$$

The initially produced hydrate is dehydrated by heating under vacuum up to 180 °C.

In the laboratory, lithium hydroxide arises by the action of water on lithium or lithium oxide. The equations for these processes follow:

$$2\ Li + 2\ H_2O \rightarrow 2\ LiOH + H_2$$

$$Li_2O + H_2O \rightarrow 2\ LiOH$$

Typically, these reactions are avoided.

Although lithium carbonate is more widely used, the hydroxide is an effective precursor to lithium salts, e.g.

$$LiOH + HF \rightarrow LiF + H_2O.$$

## Gas phase chemistry

The *acidity* of LiOH has been measured in the gas phase. The oxidolithium anion, $LiO^-$, was produced by successive decarboxylation and decarbonylation of monolithium oxalate anion, $LiO(C{=}O)(C{=}O)O^-$, by collision-induced dissociation and was identified by its exact mass. The gas-phase acidity of LiOH was inferred from the experimentally determined electron affinity of LiO• and previously known heats of formation to give a value of 426 ± 2 kcal/mol. This is considerably higher than the gas-phase acidity of water (390 kcal/mol) and even exceeds that of methane (417 kcal/mol). Thus, LiOH is a very weak acid and is in fact the weakest acid yet measured in the gas phase.

## Applications

Lithium hydroxide is mainly consumed in the production of lithium greases. A popular lithium grease thickener is Lithium 12-hydroxystearate, which produces a general-purpose lubricating grease due to its high resistance to water and usefulness at a range of temperatures.

## Carbon Dioxide Scrubbing

Lithium hydroxide is used in breathing gas purification systems for spacecraft, submarines, and rebreathers to remove carbon dioxide from exhaled gas by producing lithium carbonate and water:

$$2\ LiOH{\bullet}H_2O + CO_2 \rightarrow Li_2CO_3 + 3\ H_2O$$

or

$$2\ LiOH + CO_2 \rightarrow Li_2CO_3 + H_2O$$

The latter, anhydrous hydroxide, is preferred for its lower mass and lesser water production for respirator systems in spacecraft. One gram of anhydrous lithium hydroxide can remove 450 $cm^3$ of carbon dioxide gas. The monohydrate loses its water at 100–110 °C.

## Other uses

It is used as a heat transfer medium and as a storage-battery electrolyte. It is also used in ceramics and some Portland cement formulations. Lithium hydroxide (isotopically enriched in lithium-7) is used to alkalize the reactor coolant in pressurized water reactors for corrosion control.

## Sodium Hydroxide

Sodium hydroxide, also known as lye and caustic soda, is an inorganic compound with the formula NaOH. It is a white solid ionic compound consisting of sodium cations $Na^+$and hydroxide anions $OH^-$.

Sodium hydroxide is a highly caustic base and alkali that decomposes proteins at ordinary ambient temperatures and may cause severe chemical burns. It is highly soluble in water, and readily absorbs moisture and carbon dioxide from the air. It forms a series of hydrates $NaOH{\cdot}nH_2O$. The monohydrate $NaOH{\cdot}H_2O$ crystallizes from water solutions between 12.3 and 61.8 °C. The commercially available "sodium hydroxide" is often this monohydrate, and published data may refer to it instead of the anhydrous compound. As one of the simplest hydroxides, it is frequently utilized alongside neutral water and acidic hydrochloric acid to demonstrate the pH scale to chemistry students.

Sodium hydroxide is used in many industries: in the manufacture of pulp and paper, textiles, drinking water, soaps and detergents, and as a drain cleaner. Worldwide production in 2004 was approximately 60 million tonnes, while demand was 51 million tonnes.

## Physical Properties

Pure sodium hydroxide is a colorless crystalline solid that melts at 318 °C (604 °F) without decomposition, and with a boiling point of 1,388 °C (2,530 °F). It is highly soluble in water, with a lower solubility in polar solvents such as ethanol and methanol. NaOH is insoluble in ether and other non-polar solvents.

Similar to the hydration of sulfuric acid, dissolution of solid sodium hydroxide in water is a highly exothermic reaction where a large amount of heat is liberated, posing a threat to safety through the possibility of splashing. The resulting solution is usually colorless and odorless. As with other alkaline solutions, it feels slippery with skin contact due to the process of saponification that occurs between NaOH and natural skin oils.

## Viscosity

Sodium hydroxide, NaOH, as a fluid solution, demonstrates a characteristic viscosity, 78 mPa·s that is much greater than water (1.0 mPa·s) and near that of olive oil (85 mPa·s) at room temperature. The viscosity of NaOH, as with any chemical, is inversely related to its service temperature; meaning viscosity decreases as temperature increases, and vice versa. The viscosity of sodium hydroxide plays a direct role in its application as well as its storage.

## Hydrates

Sodium hydroxide can form several hydrates $NaOH \cdot nH_2O$, which result in a complex solubility. The known hydrates and the approximate ranges of temperature and concentration (mass percent of NaOH) of their saturated water solutions are:

- Heptahydrate, $NaOH \cdot 7H_2O$: from −28 °C (18.8%) to −24 °C (22.2%).
- Pentahydrate, $NaOH \cdot 5H_2O$: from −24 °C (22.2%) to −17.7 (24.8%).
- Tetrahydrate, $NaOH \cdot 4H_2O$, α form: from −17.7 (24.8%) to +5.4 °C (32.5%).
- Tetrahydrate, $NaOH \cdot 4H_2O$, β form: metastable.
- Trihemihydrate, $NaOH \cdot 3.5H_2O$: from +5.4 °C (32.5%) to +15.38 °C (38.8%) and then to +5.0 °C (45.7%).
- Trihydrate, $NaOH \cdot 3H_2O$: metastable.
- Dihydrate, $NaOH \cdot 2H_2O$: from +5.0 °C (45.7%) to +12.3 °C (51%).
- Monohydrate, $NaOH \cdot H_2O$: from +12.3 °C (51%) to 65.10 °C (69%) then to 62.63 °C (73.1%).

Early reports refer to hydrates with $n = 0.5$ or $n = 2/3$, but later careful investigations failed to confirm their existence.

The only hydrates with stable melting points are $NaOH \cdot H_2O$ (65.10 °C) and $NaOH \cdot 3.5H_2O$ (15.38 °C). The other hydrates, except the metastable ones $NaOH \cdot 3H_2O$ and $NaOH \cdot 4H_2O$ (β) can be crystallized from solutions of the proper composition, as listed above. However, solutions of NaOH can be easily supercooled by many degrees, which allows the formation of hydrates (including the metastable ones) from solutions with different concentrations.

For example, when a solution of NaOH and water with 1:2 mole ratio (52.6% NaOH by mass) is cooled, the monohydrate normally starts to crystallize (at about 22 °C) before the dihydrate. However, the solution can easily be supercooled down to −15 °C, at which point it may quickly crystallize as the dihydrate. When heated, the solid dihydrate might melt directly into a solution at 13.35 °C; however, once the temperature exceeds 12.58 °C. it often decomposes into solid monohydrate and a liquid solution. Even the $n = 3.5$ hydrate is difficult to crystallize, because the solution supercools so much that other hydrates become more stable.

A hot water solution containing 73.1% (mass) of NaOH is an eutectic that solidifies at about 62.63 °C as an intimate mix of anhydrous and monohydrate crystals.

A second stable eutectic composition is 45.4% (mass) of NaOH, that solidifies at about 4.9 °C into a mixture of crystals of the dihydrate and of the 3.5-hydrate.

The third stable eutectic has 18.4% (mass) of NaOH. It solidifies at about −28.7 °C as a mixture of water ice and the heptahydrate $NaOH \cdot 7H_2O$.

When solutions with less than 18.4% NaOH are cooled, water ice crystallizes first, leaving the NaOH in solution.

The α form of the tetrahydrate has density 1.33 g/cm³. It melts congruously at 7.55 °C into a liquid with 35.7% NaOH and density 1.392 g/cm³, and therefore floats on it like ice on water. However, at about 4.9 °C it may instead melt incongruously into a mixture of solid $NaOH \cdot 3.5H_2O$ and a liquid solution.

The β form of the tetrahydrate is metastable, and often transforms spontaneously to the α form when cooled below −20 °C. Once initiated, the exothermic transformation is complete in a few minutes, with a 6.5% increase in volume of the solid. The β form can be crystallized from supercooled solutions at −26 °C, and melts partially at −1.83 °C.

The "sodium hydroxide" of commerce is often the monohydrate (density 1.829 g/cm³). Physical data in technical literature may refer to this form, rather than the anhydrous compound.

## Crystal Structure

NaOH and its monohydrate form orthorhombic crystals with the space groups Cmcm (oS8) and Pbca (oP24), respectively. The monohydrate cell dimensions are a = 1.1825, b = 0.6213, c = 0.6069 nm. The atoms are arranged in a hydrargillite-like layer structure /O Na O O Na O/ Each sodium atom is surrounded by six oxygen atoms, three each from hydroxyl anions $HO^-$ and three from water molecules. The hydrogen atoms of the hydroxyls form strong bonds with oxygen atoms within each O layer. Adjacent O layers are held together by hydrogen bonds between water molecules.

## Chemical Properties

### Reaction with Acids

Sodium hydroxide reacts with protic acids to produce water and the corresponding salts. For example, when sodium hydroxide reacts with hydrochloric acid, sodium chloride is formed:

$$NaOH(aq) + HCl(aq) \rightarrow NaCl(aq) + H_2O(l)$$

In general, such neutralization reactions are represented by one simple net ionic equation:

$$OH^-(aq) + H^+(aq) \rightarrow H_2O(l)$$

This type of reaction with a strong acid releases heat, and hence is exothermic. Such acid-base reactions can also be used for titrations. However, sodium hydroxide is not used as a primary standard because it is hygroscopic and absorbs carbon dioxide from air.

### Reaction with Acidic Oxides

Sodium hydroxide also reacts with acidic oxides, such as sulfur dioxide. Such reactions are often used to "scrub" harmful acidic gases (like $SO_2$ and $H_2S$) produced in the burning of coal and thus prevent their release into the atmosphere. For example,

$$2\,NaOH + SO_2 \rightarrow Na_2SO_3 + H_2O$$

## Reaction with Metals and Oxides

Glass reacts slowly with aqueous sodium hydroxide solutions at ambient temperatures to form soluble silicates. Because of this, glass joints and stopcocks exposed to sodium hydroxide have a tendency to "freeze". Flasks and glass-lined chemical reactors are damaged by long exposure to hot sodium hydroxide, which also frosts the glass. Sodium hydroxide does not attack iron at room temperatures, since iron does not have amphoteric properties (i.e., it only dissolves in acid, not base). Nevertheless, at high temperatures (e.g. above 500 °C), iron can react endothermically with sodium hydroxide to form iron(III) oxide, sodium metal, and hydrogen gas. This is due to the lower enthalpy of formation of iron(III) oxide (−824.2kJ/mol compared to sodium hydroxide (500kJ/mol), thus the reaction is thermodynamically favorable, although its endothermic nature indicates non-spontaneity. Consider the following reaction between molten sodium hydroxide and finely divided iron filings:

$$4\ Fe + 6\ NaOH \rightarrow 2\ Fe_2O_3 + 6\ Na + 3\ H_2$$

A few transition metals, however, may react vigorously with sodium hydroxide.

In 1986, an aluminium road tanker in the UK was mistakenly used to transport 25% sodium hydroxide solution, causing pressurization of the contents and damage to the tanker. The pressurization was due to the hydrogen gas which is produced in the reaction between sodium hydroxide and aluminium:

$$2\ Al + 2\ NaOH + 6\ H_2O \rightarrow 2\ NaAl(OH)_4 + 3\ H_2$$

## Precipitant

Unlike sodium hydroxide, the hydroxides of most transition metals are insoluble, and therefore sodium hydroxide can be used to precipitate transition metal hydroxides. The following colours are observed: blue-copper, green-iron(II), yellow/brown-iron(III). Zinc and lead salts dissolve in excess sodium hydroxide to give a clear solution of $Na_2ZnO_2$ or $Na_2PbO_2$.

Aluminium hydroxide is used as a gelatinous flocculant to filter out particulate matter in water treatment. Aluminium hydroxide is prepared at the treatment plant from aluminium sulfate by reacting it with sodium hydroxide or bicarbonate.

$$Al_2(SO_4)_3 + 6\ NaOH \rightarrow 2\ Al(OH)_3 + 3\ Na_2SO_4$$

$$Al_2(SO_4)_3 + 6\ NaHCO_3 \rightarrow 2\ Al(OH)_3 + 3\ Na_2SO_4 + 6\ CO_2$$

## Saponification

Sodium hydroxide can be used for the base-driven hydrolysis of esters (as in saponification), amides and alkyl halides. However, the limited solubility of sodium hydroxide in organic solvents means that the more soluble potassium hydroxide (KOH) is often preferred. Touching sodium hydroxide solution with the bare hands, while not recommended, produces a slippery feeling. This happens because oils on the skin such as sebum are converted to soap. Despite solubility in propylene glycol it is unlikely to replace water in saponificaction due to propylene glycol primary reaction with fat before reaction between sodium hydroxide and fat.

## Production

Sodium hydroxide is industrially produced as a 50% solution by variations of the electrolytic chloralkali process. Chlorine gas is also produced in this process. Solid sodium hydroxide is obtained from this solution by the evaporation of water. Solid sodium hydroxide is most commonly sold as flakes, prills, and cast blocks.

In 2004, world production was estimated at 60 million dry metric tonnes of sodium hydroxide, and demand was estimated at 51 million tonnes. In 1998, total world production was around 45 million tonnes. North America and Asia each contributed around 14 million tonnes, while Europe produced around 10 million tonnes. In the United States, the major producer of sodium hydroxide is the Dow Chemical Company, which has annual production around 3.7 million tonnes from sites at Freeport, Texas, and Plaquemine, Louisiana. Other major US producers include Oxychem, Westlake, Olin, Shintek and Formosa. All of these companies use the chloralkali process.

Historically, sodium hydroxide was produced by treating sodium carbonate with calcium hydroxide in a metathesis reaction. (Sodium hydroxide is soluble while calcium carbonate is not.) This process was called causticizing.

$$Ca(OH)_2(aq) + Na_2CO_3(s) \rightarrow CaCO_3 \downarrow + 2\,NaOH(aq)$$

This process was superseded by the Solvay process in the late 19th century, which was in turn supplanted by the chloralkali process which we use today.

Sodium hydroxide is also produced by combining pure sodium metal with water. The byproducts are hydrogen gas and heat, often resulting in a flame, making this a common demonstration of the reactivity of alkali metals in academic environments; however, it is not commercially viable, as the isolation of sodium metal is typically performed by reduction or electrolysis of sodium compounds including sodium hydroxide.

## Uses

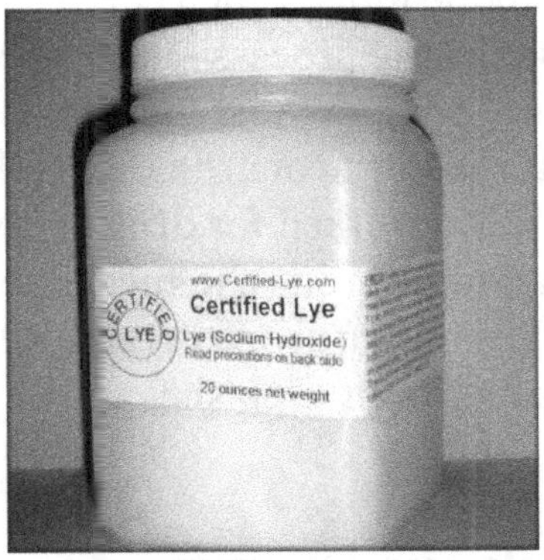


Canister of sodium hydroxide.

Sodium hydroxide is a popular strong base used in industry. Around 56% of sodium hydroxide produced is used by industry, 25% of which is used in the paper industry. Sodium hydroxide is also used in the manufacture of sodium salts and detergents, pH regulation, and organic synthesis. It is used in the Bayer process of aluminium production. In bulk, it is most often handled as an aqueous solution, since solutions are cheaper and easier to handle.

Sodium hydroxide is used in many scenarios where it is desirable to increase the alkalinity of a mixture, or to neutralize acids.

For example, in the petroleum industry, sodium hydroxide is used as an additive in drilling mud to increase alkalinity in bentonite mud systems, to increase the mud viscosity, and to neutralize any acid gas (such as hydrogen sulfide and carbon dioxide) which may be encountered in the geological formation as drilling progresses.

Poor quality crude oil can be treated with sodium hydroxide to remove sulfurous impurities in a process known as *caustic washing*. As above, sodium hydroxide reacts with weak acids such as hydrogen sulfide and mercaptans to yield non-volatile sodium salts, which can be removed. The waste which is formed is toxic and difficult to deal with, and the process is banned in many countries because of this. In 2006, Trafigura used the process and then dumped the waste in Africa.

## Chemical Pulping

Sodium hydroxide is also widely used in pulping of wood for making paper or regenerated fibers. Along with sodium sulfide, sodium hydroxide is a key component of the white liquor solution used to separate lignin from cellulose fibers in the kraft process. It also plays a key role in several later stages of the process of bleaching the brown pulp resulting from the pulping process. These stages include oxygen delignification, oxidative extraction, and simple extraction, all of which require a strong alkaline environment with a pH > 10.5 at the end of the stages.

## Tissue Digestion

In a similar fashion, sodium hydroxide is used to digest tissues, as in a process that was used with farm animals at one time. This process involved placing a carcass into a sealed chamber, then adding a mixture of sodium hydroxide and water (which breaks the chemical bonds that keep the flesh intact). This eventually turns the body into a liquid with coffee-like appearance, and the only solid that remains are bone hulls, which could be crushed between one's fingertips. Sodium hydroxide is frequently used in the process of decomposing roadkill dumped in landfills by animal disposal contractors. Due to its low cost and availability, it has been used to dispose of corpses by criminals. Italian serial killer Leonarda Cianciulli used this chemical to turn dead bodies into soap. In Mexico, a man who worked for drug cartels admitted disposing of over 300 bodies with it. Sodium hydroxide is a dangerous chemical due to its ability to hydrolyze protein. If a dilute solution is spilled on the skin, burns may result if the area is not washed thoroughly and for several minutes with running water. Splashes in the eye can be more serious and can lead to blindness.

## Dissolving Amphoteric Metals and Compounds

Strong bases attack aluminium. Sodium hydroxide reacts with aluminium and water to release hydrogen gas. The aluminium takes the oxygen atom from sodium hydroxide, which in turn takes the oxygen atom from the water, and releases the two hydrogen atoms, The reaction thus produces hydrogen gas and sodium aluminate. In this reaction, sodium hydroxide acts as an agent to make the solution alkaline, which aluminium can dissolve in. This reaction can be useful in etching, removing anodizing, or converting a polished surface to a satin-like finish, but without further passivation such as anodizing or alodining the surface may become degraded, either under normal use or in severe atmospheric conditions.

In the Bayer process, sodium hydroxide is used in the refining of alumina containing ores (bauxite) to produce alumina (aluminium oxide) which is the raw material used to produce aluminium metal via the electrolytic Hall-Héroult process. Since the alumina is amphoteric, it dissolves in the sodium hydroxide, leaving impurities less soluble at high pH such as iron oxides behind in the form of a highly alkaline red mud.

Other amphoteric metals are zinc and lead which dissolve in concentrated sodium hydroxide solutions to give sodium zincate and sodium plumbate respectively.

## Esterification and Transesterification Reagent

Sodium hydroxide is traditionally used in soap making (cold process soap, saponification). It was made in the nineteenth century for a hard surface rather than liquid product because it was easier to store and transport.

For the manufacture of biodiesel, sodium hydroxide is used as a catalyst for the transesterification of methanol and triglycerides. This only works with anhydrous sodium hydroxide, because combined with water the fat would turn into soap, which would be tainted with methanol. NaOH is used more often than potassium hydroxide because it is cheaper and a smaller quantity is needed.

## Food Preparation

Food uses of sodium hydroxide include washing or chemical peeling of fruits and vegetables, chocolate and cocoa processing, caramel coloring production, poultry scalding, soft drink processing, and thickening ice cream. Olives are often soaked in sodium hydroxide for softening; Pretzels and German lye rolls are glazed with a sodium hydroxide solution before baking to make them crisp. Owing to the difficulty in obtaining food grade sodium hydroxide in small quantities for home use, sodium carbonate is often used in place of sodium hydroxide. It is known as E number E524.

Specific foods processed with sodium hydroxide include:

- German pretzels are poached in a boiling sodium carbonate solution or cold sodium hydroxide solution before baking, which contributes to their unique crust.
- Lye-water is an essential ingredient in the crust of the traditional baked Chinese moon cakes.
- Most yellow coloured Chinese noodles are made with lye-water but are commonly mistaken for containing egg.
- One variety of zongzi uses lye water to impart a sweet flavor.
- Sodium hydroxide is also the chemical that causes gelling of egg whites in the production of Century eggs.

- Some methods of preparing olives involve subjecting them to a lye-based brine.
- The Filipino dessert (*kakanin*) called *kutsinta* uses a small quantity of lye water to help give the rice flour batter a jelly like consistency. A similar process is also used in the kakanin known as *pitsi-pitsi* or *pichi-pichi* except that the mixture uses grated cassava instead of rice flour.
- The Norwegian dish known as lutefisk (from *lutfisk*, "lye fish").
- Bagels are often boiled in a lye solution before baking, contributing to their shiny crust.
- Hominy is dried maize (corn) kernels reconstituted by soaking in lye-water. These expand considerably in size and may be further processed by frying to make corn nuts or by drying and grinding to make grits. Hominy is used to create Masa, a popular flour used in Mexican cuisine to make Corn tortillas and tamales. Nixtamal is similar, but uses calcium hydroxide instead of sodium hydroxide.

## Cleaning Agent

Sodium hydroxide is frequently used as an industrial cleaning agent where it is often called "caustic". It is added to water, heated, and then used to clean process equipment, storage tanks, etc. It can dissolve grease, oils, fats and protein-based deposits. It is also used for cleaning waste discharge pipes under sinks and drains in domestic properties. Surfactants can be added to the sodium hydroxide solution in order to stabilize dissolved substances and thus prevent redeposition. A sodium hydroxide soak solution is used as a powerful degreaser on stainless steel and glass bakeware. It is also a common ingredient in oven cleaners.

A common use of sodium hydroxide is in the production of parts washer detergents. Parts washer detergents based on sodium hydroxide are some of the most aggressive parts washer cleaning chemicals. The sodium hydroxide-based detergents include surfactants, rust inhibitors and defoamers. A parts washer heats water and the detergent in a closed cabinet and then sprays the heated sodium hydroxide and hot water at pressure against dirty parts for degreasing applications. Sodium hydroxide used in this manner replaced many solvent-based systems in the early 1990s when trichloroethane was outlawed by the Montreal Protocol. Water and sodium hydroxide detergent-based parts washers are considered to be an environmental improvement over the solvent-based cleaning methods.

Hardware stores grade sodium hydroxide to be used as a type of drain cleaner.

Paint stripping with caustic soda.

Sodium hydroxide is used in the home as a type of drain opener to unblock clogged drains, usually in the form of a dry crystal or as a thick liquid gel. The alkali dissolves greases to produce water soluble products. It also hydrolyzes the proteins such as those found in hair which may block water pipes. These reactions are sped by the heat generated when sodium hydroxide and the other chemical components of the cleaner dissolve in water. Such alkaline drain cleaners and their acidic versions are highly corrosive and should be handled with great caution.

Sodium hydroxide is used in some relaxers to straighten hair. However, because of the high incidence and intensity of chemical burns, manufacturers of chemical relaxers use other alkaline chemicals in preparations available to average consumers. Sodium hydroxide relaxers are still available, but they are used mostly by professionals.

A solution of sodium hydroxide in water was traditionally used as the most common paint stripper on wooden objects. Its use has become less common, because it can damage the wood surface, raising the grain and staining the colour.

## Water Treatment

Sodium hydroxide is sometimes used during water purification to raise the pH of water supplies. Increased pH makes the water less corrosive to plumbing and reduces the amount of lead, copper and other toxic metals that can dissolve into drinking water.

## Historical uses

Sodium hydroxide has been used for detection of carbon monoxide poisoning, with blood samples of such patients turning to a vermilion color upon the addition of a few drops of sodium hydroxide. Today, carbon monoxide poisoning can be detected by CO oximetry.

## In Cement Mixes, Mortars, Concrete, Grouts

Sodium hydroxide is used in some cement mix plasticisers. This helps homogenise cement mixes, preventing segregation of sands and cement, decreases the amount of water required in a mix and increases workability of the cement product, be it mortar, render or concrete.

## Summer-winter Heat Storage

EMPA researchers are experimenting with concentrated sodium hydroxide (NaOH) as the thermal storage or seasonal reservoir medium for domestic space-heating. If water is added to solid or concentrated sodium hydroxide (NaOH), heat is released. The dilution is exothermic – chemical energy is released in the form of heat. Conversely, by applying heat energy into a dilute sodium hydroxide solution the water will evaporate so that the solution becomes more concentrated and thus stores the supplied heat as latent chemical energy.

## Storage

Careful storage is needed when handling sodium hydroxide for use, especially bulk volumes. Following proper NaOH storage guidelines and maintaining worker/environment safety is always recommended given the chemical's burn hazard. Sodium hydroxide is often stored in bottles—as in laboratories, small-scale use—within intermediate bulk containers—medium volume containers for cargo handling and transport—or within large stationary storage tanks with volumes up to 100,000 gallons—as in manufacturing or waste water plants with extensive NaOH use. Common materials that are compatible with sodium hydroxide and often utilized for NaOH storage include: polyethylene (HDPE, usual, XLPE, less common), carbon steel, polyvinyl chloride (PVC), stainless steel, and fiberglass reinforced plastic (FRP, with a resistant liner).

## Potassium Hydroxide

Potassium hydroxide is an inorganic compound with the formula KOH, and is commonly called caustic potash.

Along with sodium hydroxide (NaOH), this colorless solid is a prototypical strong base. It has many industrial and niche applications, most of which exploit its caustic nature and its reactivity toward acids. An estimated 700,000 to 800,000 tonnes were produced in 2005. About 100 times more NaOH than KOH is produced annually. KOH is noteworthy as the precursor to most soft and liquid soaps, as well as numerous potassium-containing chemicals. It is a white solid that is dangerously corrosive. Most commercial samples are ca. 90% pure, the remainder being water and carbonates.

## Properties and Structure

Potassium hydroxide is usually sold as translucent pellets, which become tacky in air because KOH is hygroscopic. Consequently, KOH typically contains varying amounts of water. Its dissolution in water is strongly exothermic. Concentrated aqueous solutions are sometimes called potassium lyes. Even at high temperatures, solid KOH does not dehydrate readily.

## Structure

At higher temperatures, solid KOH crystallizes in the NaCl crystal structure. The OH group is either rapidly or randomly disordered so that the $OH^-$ group is effectively a spherical anion of radius 1.53 Å (between $Cl^-$ and $F^-$ in size). At room temperature, the $OH^-$ groups are ordered and the environment about the $K^+$ centers is distorted, with $K^+$_$OH^-$ distances ranging from 2.69 to 3.15 Å, depending on the orientation of the OH group. KOH forms a series of crystalline hydrates, namely the monohydrate $KOH \cdot H_2O$, the dihydrate $KOH \cdot 2H_2O$ and the tetrahydrate $KOH \cdot 4H_2O$.

### Thermal Stability

Like NaOH, KOH exhibits high thermal stability. The gaseous species is dimeric. Because of its high stability and relatively low melting point, it is often melt-cast as pellets or rods, forms that have low surface area and convenient handling properties.

## Reactions

### Basicity, Solubility and Desiccating Properties

About 121 g of KOH dissolve in 100 mL water at room temperature, which contrasts with 100 g/100 mL for NaOH. Thus on a molar basis, KOH is slightly less soluble than NaOH. Lower molecular-weight alcohols such as methanol, ethanol, and propanols are also excellent solvents. They participate in an acid-base equilibrium. In the case of methanol the potassium methoxide (methylate) forms:

$$KOH + CH_3OH \rightleftharpoons CH_3OK + H_2O$$

Because of its high affinity for water, KOH serves as a desiccant in the laboratory. It is often used to dry basic solvents, especially amines and pyridines.

### As a Nucleophile in Organic Chemistry

KOH, like NaOH, serves as a source of $OH^-$, a highly nucleophilic anion that attacks polar bonds in both inorganic and organic materials. Aqueous KOH saponifies esters:

$$KOH + RCOOR' \rightarrow RCOOK + R'OH$$

When R is a long chain, the product is called a potassium soap. This reaction is manifested by the "greasy" feel that KOH gives when touched — fats on the skin are rapidly converted to soap and glycerol.

Molten KOH is used to displace halides and other leaving groups. The reaction is especially useful for aromatic reagents to give the corresponding phenols.

### Reactions with Inorganic Compounds

Complementary to its reactivity toward acids, KOH attacks oxides. Thus, $SiO_2$ is attacked by KOH to give soluble potassium silicates. KOH reacts with carbon dioxide to give bicarbonate:

$$KOH + CO_2 \rightarrow KHCO_3$$

## Manufacture

Historically, KOH was made by adding potassium carbonate to a strong solution of calcium hydroxide (slaked lime) The salt metathesis reaction results in precipitation of solid calcium carbonate, leaving potassium hydroxide in solution:

$$Ca(OH)_2 + K_2CO_3 \rightarrow CaCO_3 + 2\ KOH$$

Filtering off the precipitated calcium carbonate and boiling down the solution gives potassium hydroxide ("calcinated or caustic potash"). This method of producing potassium hydroxide remained dominant until the late 19th century, when it was largely replaced by the current method of electrolysis of potassium chloride solutions. The method is analogous to the manufacture of sodium hydroxide.

$$2\ KCl + 2\ H_2O \rightarrow 2\ KOH + Cl_2 + H_2$$

Hydrogen gas forms as a byproduct on the cathode; concurrently, an anodic oxidation of the chloride ion takes place, forming chlorine gas as a byproduct. Separation of the anodic and cathodic spaces in the electrolysis cell is essential for this process.

## Uses

KOH and NaOH can be used interchangeably for a number of applications, although in industry, NaOH is preferred because of its lower cost.

## Precursor to other Potassium Compounds

Many potassium salts are prepared by neutralization reactions involving KOH. The potassium salts of carbonate, cyanide, permanganate, phosphate, and various silicates are prepared by treating either the oxides or the acids with KOH. The high solubility of potassium phosphate is desirable in fertilizers.

## Manufacture of Soft Soaps

The saponification of fats with KOH is used to prepare the corresponding "potassium soaps", which are softer than the more common sodium hydroxide-derived soaps. Because of their softness and greater solubility, potassium soaps require less water to liquefy, and can thus contain more cleaning agent than liquefied sodium soaps.

## As an Electrolyte

Potassium carbonate, formed from the hydroxide solution leaking from an alkaline battery.

Aqueous potassium hydroxide is employed as the electrolyte in alkaline batteries based on nickel-cadmium, nickel-hydrogen, and manganese dioxide-zinc. Potassium hydroxide is preferred over sodium hydroxide because its solutions are more conductive. The nickel–metal hydride batteries in the Toyota Prius use a mixture of potassium hydroxide and sodium hydroxide. Nickel–iron batteries also use potassium hydroxide electrolyte.

## Food Industry

In food products, potassium hydroxide acts as a food thickener, pH control agent and food stabilizer. The FDA considers it (as a direct human food ingredient) as generally safe when combined with "good" manufacturing practice conditions of use. It is known in the E number system as E525.

## Niche Applications

Like sodium hydroxide, potassium hydroxide attracts numerous specialized applications, virtually all of which rely on its properties as a strong chemical base with its consequent ability to degrade many materials. For example, in a process commonly referred to as "chemical cremation" or "resomation", potassium hydroxide hastens the decomposition of soft tissues, both animal and human, to leave behind only the bones and other hard tissues. Entomologists wishing to study the fine structure of insect anatomy may use a 10% aqueous solution of KOH to apply this process.

In chemical synthesis, the choice between the use of KOH and the use of NaOH is guided by the solubility or keeping quality of the resulting salt.

The corrosive properties of potassium hydroxide make it a useful ingredient in agents and preparations that clean and disinfect surfaces and materials that can themselves resist corrosion by KOH. KOH is also used for semiconductor chip fabrication. Potassium hydroxide is often the main active ingredient in chemical "cuticle removers" used in manicure treatments.

Because aggressive bases like KOH damage the cuticle of the hair shaft, potassium hydroxide is used to chemically assist the removal of hair from animal hides. The hides are soaked for several hours in a solution of KOH and water to prepare them for the unhairing stage of the tanning process. This same effect is also used to weaken human hair in preparation for shaving. Preshave products and some shave creams contain potassium hydroxide to force open the hair cuticle and to act as a hygroscopic agent to attract and force water into the hair shaft, causing further damage to the hair. In this weakened state, the hair is more easily cut by a razor blade.

Potassium hydroxide is used to identify some species of fungi. A 3–5% aqueous solution of KOH is applied to the flesh of a mushroom and the researcher notes whether or not the color of the flesh changes. Certain species of gilled mushrooms, boletes, polypores, and lichens are identifiable based on this color-change reaction

## Calcium Hydroxide

Calcium hydroxide (traditionally called slaked lime) is an inorganic compound with the chemical formula $Ca(OH)_2$. It is a colorless crystal or white powder and is produced when *quicklime* (calcium oxide) is mixed, or slaked with water. It has many names including hydrated lime, caustic lime, builders' lime, slack lime, cal, or pickling lime. Calcium hydroxide is used in many applications, including food preparation, where it has been identified as E number E526. Limewater is the common name for a saturated solution of calcium hydroxide.

## Properties

Calcium hydroxide is relatively insoluble in water, with a solubility product $K_{sp}$ of $5.5 \times 10^{-6}$. It is large enough that its solutions are basic according to the following reaction:

$$Ca(OH)_2 \rightarrow Ca^{2+} + 2\ OH^-$$

At ambient temperature, calcium hydroxide (portlandite) dissolves in pure water to produce an alkaline solution with a pH of about 12.4. Calcium hydroxide solutions can cause chemical burns. At high pH value, its solubility drastically decreases. This behavior is relevant to cement pastes. Aqueous solutions of calcium hydroxide are called limewater and are medium strength bases that reacts with acids and can attack some metals such as aluminium (amphoteric hydroxide dissolving at high pH) while protecting other metals from corrosion such as iron and steel by passivation of their surface. Limewater turns milky in the presence of carbon dioxide due to formation of calcium carbonate, a process called carbonatation:for example lime water,

$$Ca(OH)_2 + CO_2 \rightarrow CaCO_3 + H_2O$$

When heated to 512 °C, the partial pressure of water in equilibrium with calcium hydroxide reaches 101 kPa (normal atmospheric pressure), which decomposes calcium hydroxide into calcium oxide and water.

$$Ca(OH)_2 \rightarrow CaO + H_2O$$

## Structure, Preparation, Occurrence

SEM image of fractured hardened cement paste, showing plates of calcium hydroxide and needles of ettringite (micron scale).

Calcium hydroxide adopts a polymeric structure, as do all metal hydroxides. The structure is identical to that of $Mg(OH)_2$ (*brucite structure*); i.e., the cadmium iodide motif. Strong hydrogen bonds exist between the layers.

Calcium hydroxide is produced commercially by treating lime with water:

$$CaO + H_2O \rightarrow Ca(OH)_2$$

In the laboratory it can be prepared by mixing aqueous solutions of calcium chloride and sodium hydroxide. The mineral form, portlandite, is relatively rare but can be found in some volcanic, plutonic, and metamorphic rocks. It has also been known to arise in burning coal dumps.

The positively charged ionized species $CaOH^+$ has been detected in the atmosphere of S-type stars.

## Retrograde Solubility

The solubility of calcium hydroxide (portlandite) at 70 °C is about half of its value at 25 °C. The reason for this rather uncommon phenomenon is that the dissolution of calcium hydroxide in water is an exothermic process, and also adheres to Le Chatelier's principle. A lowering of temperature thus favours the elimination of the heat liberated through the process of dissolution and increases the equilibrium constant of dissolution of $Ca(OH)_2$, and so increase its solubility at low temperature. This counter-intuitive temperature dependence of the solubility is referred to as "retrograde" or "inverse" solubility. The variably hydrated phases of calcium sulfate (gypsum, bassanite and anhydrite) also exhibit a retrograde solubility for the same reason because their dissolution reactions are exothermic.

## Uses

One significant application of calcium hydroxide is as a flocculant, in water and sewage treatment. It forms a fluffy charged solid that aids in the removal of smaller particles from water, resulting in a clearer product. This application is enabled by the low cost and low toxicity of calcium hydroxide. It is also used in fresh water treatment for raising the pH of the water so that pipes will not corrode where the base water is acidic, because it is self-regulating and does not raise the pH too much.

It is also used in the preparation of ammonia gas ($NH_3$), using the following reaction:

$$Ca(OH)_2 + 2NH_4Cl \rightarrow 2NH_3 + CaCl_2 + 2H_2O$$

Another large application is in the paper industry, where it is an intermediate in the reaction in the production of sodium hydroxide. This conversion is part of the *causticizing* step in the Kraft process for making pulp. In the causticizing operation, burned lime is added to *green liquor*, which is a solution primarily of sodium carbonate and sodium sulfate produced by dissolving *smelt*, which is the molten form of these chemicals from the recovery furnace.

## Food Industry

Because of its low toxicity and the mildness of its basic properties, slaked lime is widely used in the food industry:

- In USDA certified food production in plants and livestock.
- To clarify raw juice from sugarcane or sugar beets in the sugar industry.
- To process water for alcoholic beverages and soft drinks.
- Pickle cucumbers and other foods.
- To make Chinese century eggs.

- In maize preparation: removes the cellulose hull of maize kernels.
- To clear a brine of carbonates of calcium and magnesium in the manufacture of salt for food and pharmaceutical uses.
- In fortifying (Ca supplement) fruit drinks, such as orange juice, and infant formula.
- As a digestive aid (called Choona, used in India in *paan*, a mixture of areca nuts, calcium hydroxide and a variety of seeds wrapped in betel leaves).
- As a substitute for baking soda in making *papadam*.
- In the removal of carbon dioxide from controlled atmosphere produce storage rooms.

## Native American uses

Dry untreated maize (left), and treated maize (right) after boiling in water with calcium hydroxide (15 ml, or 1 Tbsp, lime for 500 g of corn) for 15 minutes.

In Spanish, calcium hydroxide is called *cal*. Maize cooked with *cal* (nixtamalization) becomes hominy (*nixtamal*), which significantly increases the bioavailability of niacin (vitamin B3), and it is also considered tastier and easier to digest.

In chewing coca leaves, calcium hydroxide is usually chewed alongside to keep the alkaloid stimulants chemically available for absorption by the body. Similarly, Native Americans traditionally chewed tobacco leaves with calcium hydroxide derived from burnt mollusc shells to enhance the effects. It has also been used by some indigenous American tribes as an ingredient in *yopo*, a psychedelic snuff prepared from the beans of some *Anadenanthera* species.

## Asian uses

Calcium hydroxide is typically added to a bundle of areca nut and betel leaf called 'paan' to keep the alkaloid stimulants chemically available to enter the bloodstream via sublingual absorption.

It is used in making *naswar* (also known as *nass* or *niswar*), a type of dipping tobacco made from fresh tobacco leaves, calcium hydroxide (*chuna* or *soon*), and wood ash. It is consumed most in the Pathan diaspora, Afghanistan, Pakistan, India and Bangladesh. Villagers also use calcium hydroxide to paint their mud houses in Afghanistan, Pakistan and India.

# Sodium Hydrogen Carbonate

Sodium hydrogen carbonate is a weakly basic inorganic compound, commonly called as sodium bicarbonate or baking soda.

Formula and structure: The chemical formula of sodium hydrogen carbonate is $NaHCO_3$, and its molar mass is 84.007 g/mol. Its chemical structure is shown below. It is an ionic compound made of sodium cations ($Na^+$) and bicarbonate anions ($HCO_3^-$).

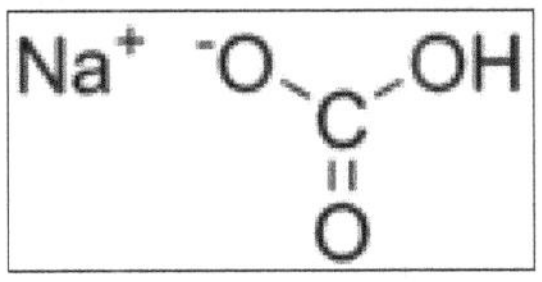

Occurrence: Sodium hydrogen carbonate occurs naturally as the mineral nahcolite, which is found in many mineral springs. This is a major source of this material.

Preparation: Sodium hydrogen carbonate is also prepared on large scales by the Solvay process, in which sodium chloride reacts with ammonia, carbon dioxide and water to give $NaHCO_3$ along with ammonium chloride salt ($NH_4Cl$).

$$NaCl + NH_3 + CO_2 + H_2O \rightarrow NaHCO_3 + NH_4Cl$$

Another main preparation method involves dissolving soda ash (sodium carbonate mineral) in water and passing carbon dioxide through the solution.

$$Na_2CO_3 + CO_2 + H_2O \rightarrow 2\ NaHCO_3$$

Physical properties: Sodium hydrogen carbonate is an odorless, white crystalline solid with a density of 2.2 g/mL. It is also commonly available as a fine white powder, with a density of 1.2 g/mL. Its melting point is 50 °C, and it is highly water soluble.

Chemical properties: $NaHCO_3$ dissolves in water to give weakly acidic carbonic acid ($H_2CO_3$) and the basic hydroxide ions, making it an overall weak base. Sodium hydrogen carbonate reacts with acids to give salts and carbonic acid, which then decomposes into carbon dioxide gas and water:

$$NaHCO_3 + HCl \rightarrow NaCl + H_2CO_3 \rightarrow H_2O + CO_2$$

It decomposes upon heating into sodium carbonate, water and carbon dioxide:

$$2\ NaHCO_3 \rightarrow Na_2CO_3 + H_2O + CO_2$$

It also reacts with bases such as sodium hydroxide to form sodium carbonate:

$$NaHCO_3 + NaOH \rightarrow Na_2CO_3 + H_2O$$

Uses: Sodium hydrogen carbonate is commonly used as an antacid, in baking powder, as odor absorber, drying agent, and in toothpastes. It is also used as a fire retardant, in fireworks, for mild disinfecting, pest control, pH balancing, paint removal and metal cleaning applications.

Health effects/safety hazards: It is not considered toxic or hazardous. However, eye or skin contact with the concentrated $NaHCO_3$ solutions can cause severe eye and skin irritation.

## References


- Haynes, William M., ed. (2011). CRC Handbook of Chemistry and Physics (92nd ed.). Boca Raton, FL: CRC Press. p. 4.46. ISBN 1439855110.
- Inorganic-compounds-examples-321, Science, examples: softschools.com, Retrieved 16 April, 2019
- Inorganic-hydroxides, en: alfa.com, Retrieved 16 April, 2019
- Zapp, Karl-Heinz (2012) “Ammonium Compounds” in Ullmann’s Encyclopedia of Industrial Chemistry. Wiley-VCH, Weinheim. doi:10.1002/14356007.a02_243
- Formulas, chemistry, sodium-hydrogen-carbonate-uses-properties-structure-formula-250: softschools.com, Retrieved 24 June, 2019

# 5
# Inorganic Reactions

The reactions which involve inorganic compounds are known as inorganic reactions. Some of the major types of inorganic reactions are combination reactions, decomposition reactions, single-replacement reactions and double-replacement reactions. This chapter has been carefully written to provide an easy understanding of these types of reactions.

Many chemical reactions can be classified as one of five basic types. Having a thorough understanding of these types of reactions will be useful for predicting the products of an unknown reaction. The five basic types of chemical reactions are combination, decomposition, single-replacement, double-replacement, and combustion. Analyzing the reactants and products of a given reaction will allow you to place it into one of these categories. Some reactions will fit into more than one category. Identifying a reaction as one of these types does not preclude it from also being an oxidation-reduction (redox) reaction. For example, many combination reactions can also be classified as a redox reaction.

## Combination Reactions

A combination reaction is a general category of chemical reactions. It may be defined as a chemical reaction in which two or more substances combine to form a single substance under suitable conditions.

Combination reactions are also known as synthesis, because in these reactions new substances are synthesized.

The general equation that describes a combination reaction is:

### Types of Combination Reaction

Combination reactions are of three types:

$$2Na(s) + CI_2(g) \longrightarrow 2NaCI(g)$$

Similarly, non-metals may react with highly active metals to form covalent compounds. Example, sulphur reacts with oxygen gas to form gaseous sulphur dioxide.

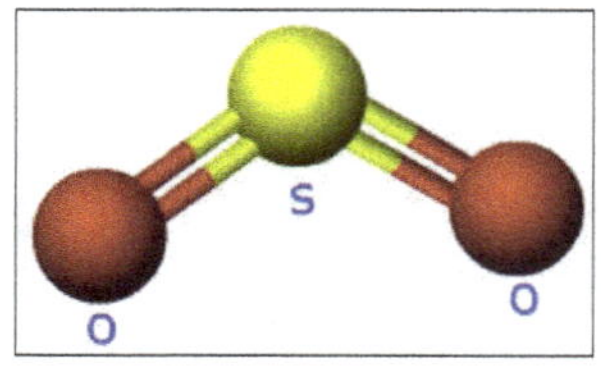


$$S(s) + O_2(g) \longrightarrow 2SO_2(g)$$

- Reaction between two or more elements: An example of this type of combination reaction is the reaction between a metal and a non-metal. Most metals react with non-metals to form ionic compounds. A good example of this would be:

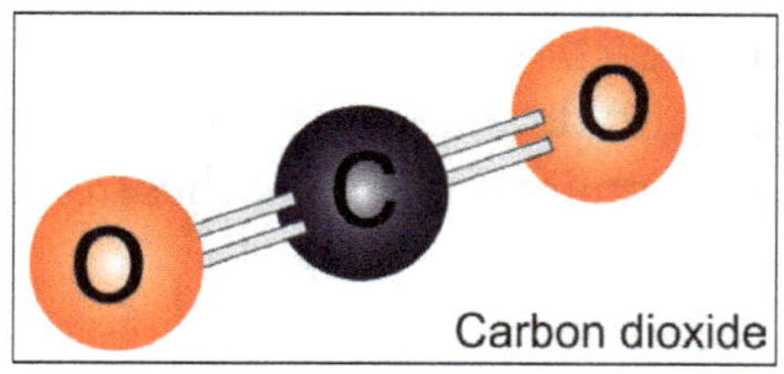


$$CaO(s) + CO(g) \longrightarrow CaCO^3(S)$$

- Reaction between elements and compounds: An element and a compound react to form another compound. Example, carbon monoxide reacts with oxygen gas to form carbon dioxide.
- Reaction between two compounds: Two compounds react with each other to form a new compound. Example, calcium oxide (quick lime) reacts with carbon dioxide gas to form calcium carbonate (lime stone).

$$CaO(s) + CO_2(g) \longrightarrow CaCO_3(S)$$

## Most Combination Reactions are Exothermic in Nature

Combination reactions involve the formation of new bonds and this process releases a large amount of energy in the form of heat.

Let us discuss some more examples of combination reactions.

- Formation of Calcium Hydroxide: Reaction between quick lime (Calcium oxide, CaO) and water is a combination reaction.

In this reaction, quick lime reacts with water to form slaked lime (calcium hydroxide, Ca(OH)2). The reaction between quick lime and water is highly vigorous as well as exothermic.

$$\underset{\text{Quick time}}{CaO(s)} + H_2O(I) \longrightarrow \underset{\text{Slaked time}}{Ca(OH)_2(s)}$$

- Combustion Reactions: Combustion of coal and combustion of hydrogen are examples of combination reactions.

Coal burns in air to form carbon dioxide gas.

$$C(s)+O_2(g) \longrightarrow CO_2(g)$$

Hydrogen burns in the presence of oxygen to form water in the form of steam. Upon cooling, it becomes liquid.

$$2H_2(g)+O_2(g) \longrightarrow 2H_2O(I)$$

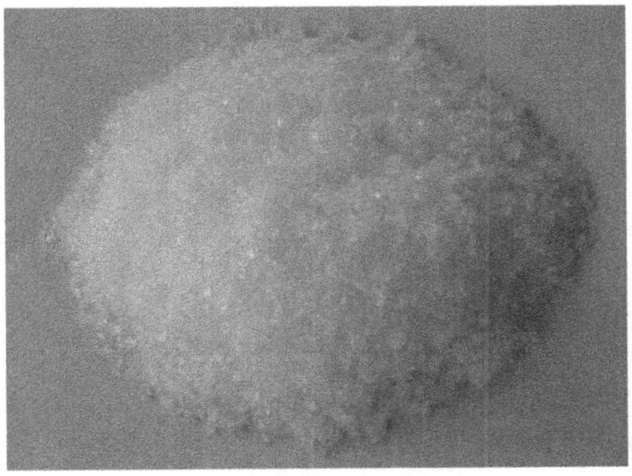

- Formation of Ammonium chloride: Ammonium chloride is formed by combining vapours of ammonia with hydrogen chloride gas. It is a white-coloured solid.

$$NH_3(g)+HCI(g) \longrightarrow NH_4CI(s)$$

- Formation of Sulphuric acid: Formation of sulphuric acid from sulphur trioxide is also a combination reaction. Sulphur trioxide on hydration forms sulphuric acid. This reaction is highly exothermic in nature.

$$SO_3(g)+H_2O(I) \longrightarrow H_2SO_4(I)$$

- Formation of ferrous sulphide: It is formed by heating fine pieces of iron with sulphur powder.

$$Fe(s)+S(s) \longrightarrow FeS(s)$$

- Manufacture of Ammonia: The manufacture of ammonia gas from nitrogen and oxygen is also a combination reaction.

$$N_2(g)+3H_2(g) \longrightarrow 2NH_3(g)$$

- Combination of sodium oxide and water: Sodium oxide combines with water to form sodium hydroxide.

$$Na_2O(s)+H_2O(I) \longrightarrow 2NaOH(s)$$

## Decomposition Reactions

A decomposition reaction can be defined as a type of chemical reaction in which under suitable conditions, one single compound splits into multiple simpler substances.

### General Reaction Mechanism

AB -> A + B

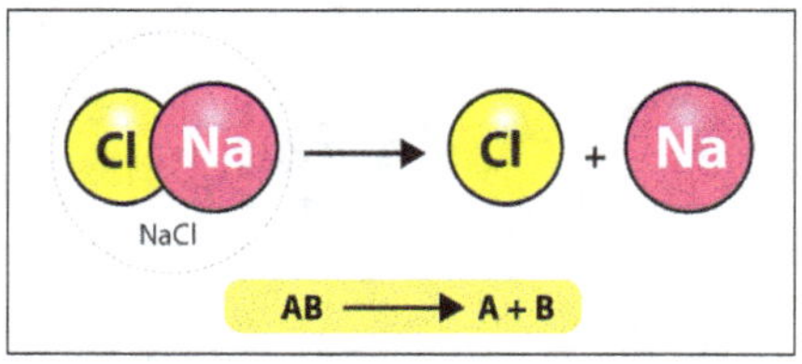


### Decomposition Reaction Examples

Decomposition reaction happens all throughout, but we often cease to see it. Let us look at some decomposition reaction examples in real life.

- $KClO_3$ decomposition reaction.
- $H_2O_2$ decomposition reaction.

Even Though the decomposition reaction concept is simple, these are some of the common doubts people have about the concept.

### Combination Reaction

We know a combination reaction to be a chemical reaction in two or more substances combine together to produce a new compound. Whereas in a decomposition reaction, the combination between constituents are broken down to form multiple products.

### Endothermic Decomposition Reaction

No, not all decomposition reactions are endothermic. A decomposition reaction can be both endothermic or exothermic, even though the latter is more predominant.

Examples:

- Decomposition of NO to $N_2$ & $O_2$ is exothermic.

- Decomposition of ozone ($O_3$) to oxygen ($O_2$) is exothermic.

## Double Decomposition Reaction

A double decomposition reaction is a type of decomposition reaction in which two constituent reactants interchange positive and negative ions and form two new compounds.

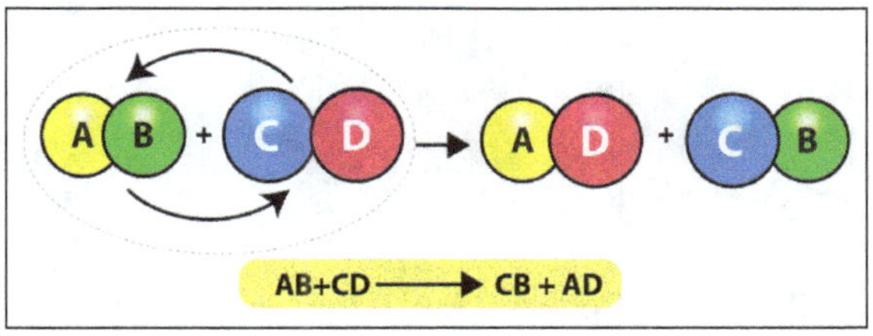


Example

$HCl(aq) + NaOH(aq) \rightarrow NaCl(aq) + HOH(l)$

## Types of Decomposition Reaction

Let us look at the three main types of decomposition reactions which we see around us:

- Thermal decomposition reaction A thermal decomposition reaction can be defined as a decomposition reaction which is activated by thermal energy. The reaction is generally endothermic as energy is required to break chemical bond in order to separate the constituent elements. Example of thermal decomposition reaction. $CaCO_3 \rightarrow CaO + CO_2$. When heated Calcium Carbonate,decomposes into Calcium Oxide and Carbon Dioxide. This process is employed in the manufacturing of quick lime which is an important component in industries.
- Electrolytic decomposition reaction. An electrolytic decomposition reaction is a type of decomposition reaction in which the activation energy for decomposition. Example of electrolytic decomposition reaction. Dissociation of water: $2H_2O \rightarrow 2H_2 + O_2$.
- Photo decomposition/photolytic decomposition/photochemical decomposition. A photo decomposition is the type of decomposition reaction in which the reactant is broken down to its constituents by photons. Example of photo decomposition reaction. $O_3 + hv \rightarrow O_2 + O(1D)$ Photolysis occurs in the atmosphere wherein ozone molecules is broken down into oxygen molecule and atomic oxygen.

## Uses of Decomposition Reaction

One major application of the decomposition is the extraction of metals from its ore. Like obtaining zinc from calamine and Sodium from sodium from sodium chloride. Other uses are:

- Manufacture of cement or calcium oxide.
- For metallurgical processes: Extraction of metals from its oxides, chloride etc.
- The relief from acid indigestion.
- Thermite welding.

# Single-replacement Reactions

Copper displaces dissolved silver from a solution. When a copper wire is dipped in a silver nitrate solution, solid silver precipitates out.

A single-displacement reaction, also known as a single-replacement reaction, is a chemical reaction in which one (or more) element(s) replaces an/other element(s) in a compound. It can be represented generically as:

$$A + B\text{-}C \rightarrow A\text{-}C + B$$

This will most often occur if A is more reactive than B, thus giving a more stable product. A and B must be either:

- Different metals (hydrogen's behavior as a cation renders it as a metal here), in which case C represents an anion; or
- Halogens, in which case C represents a cation.

In either case, when AC and BC are aqueous compounds (which is usually the case), C is a spectator ion.

In the activity or reactivity series, the metals with the highest propensity to donate their electrons to react are listed first, and the most unreactive metals are listed last. Therefore, a metal higher on the list is able to displace anything on the list below it. The order of activity for metals, from most reactive to least reactive, is: Li, K, Sr, Na, Ca, Mg, Al, Zn, Cr, Fe, Cd, Co, Ni, Sn, Pb, H, Sb, As, Bi, Cu, Hg, Ag, Pd, Pt, and Au. Similarly, the halogens with the highest propensity to acquire electrons are the most reactive. The activity series for halogens, from highest to lowest, is: F, Cl, Br, and I.

Due to the free state nature of A and B, all single displacement reactions are also oxidation-reduction reactions, where the key event is the movement of electrons from one reactant to another. When A and B are metals, A is always oxidized and B is always reduced. Since halogens prefer to gain electrons, A is reduced (from 0 to −1) and B is oxidized (from −1 to 0) when A and B represent those elements.

A and B may have different charge as ions and therefore some balancing of the equation may be necessary. For example, the reaction between silver nitrate, $AgNO_3$ (which contains an $Ag^+$ ion), and zinc, Zn, forms silver, Ag, and zinc nitrate, $Zn(NO_3)_2$ (which contains a $Zn^{2+}$ ion).

$$2AgNO_{3(aq)} + Zn_{(s)} \rightarrow 2Ag_{(s)} + Zn(NO_3)_{2(aq)}$$

All simple metal with acid reactions are single displacement reactions. For example, the reaction between magnesium, Mg, and hydrochloric acid, HCl, forms magnesium chloride, $MgCl_2$, and hydrogen, $H_2$.

$$Mg_{(s)} + 2\ HCl_{(aq)} \rightarrow MgCl_{2(aq)} + H_{2(g)}$$

## Cation Replacement

One cation replaces another. A cation is a positively charged ion or a metal. When it is written in generic symbols, it is written out like this:

$$X + YZ \rightarrow XZ + Y$$

Element X has replaced Y in compound YZ to become a new compound XZ and the free element Y. This is an oxidation–reduction reaction wherein element Y is reduced from a cation into the elemental form and element X is oxidized from the elemental form into a cation.

Some examples are:

$$Cu+2AgNO_3 \rightarrow 2Ag \downarrow +Cu(NO_3)_2$$
$$Fe+Cu(NO_3)_2 \rightarrow Fe(NO_3)_2+Cu \downarrow$$
$$Ca+2H_2O \rightarrow Ca(OH)_2+H_2 \uparrow$$
$$Zn+2HCl \rightarrow ZnC12+H2 \uparrow$$

If the reactant in elemental form is not the more reactive metal, then no reaction will occur. Some examples of this would be the reverse reactions to these:

1. $Ag + Cu(NO_3)_2 \rightarrow$ No reaction
2. $Au + HCl \rightarrow$ No reaction

## Anion Replacement

One anion replaces another. An anion is a negatively charged ion or a nonmetal. Written using generic symbol, it is:

$$A + XY \rightarrow XA + Y$$

Element A has replaced Y (in the compound XY) to form a new compound XA and the free element Y. This is an oxidation-reduction reaction wherein element A is reduced from the elemental form into an anion and element Y is oxidized from an anion into the elemental form.

Some of the few examples that involve halogens are shown here:

$$Cl_2+2NaBr \rightarrow 2NaCl+Br_2 \downarrow$$
$$Br_2+2KI \rightarrow 2KBr+I_2 \downarrow$$

Again, the less reactive halogen cannot replace the more reactive halogen:

$$I_2 + 2KBr \rightarrow \text{no reaction.}$$

## Double-replacement Reactions

A double-replacement reaction is a reaction in which the positive and negative ions of two ionic compounds exchange places to form two new compounds. The general form of a double-replacement (also called double-displacement) reaction is:

$$AB+CD \rightarrow AD+BC$$

In this reaction, A and C are positively-charged cations, while B and D are negatively-charged anions. Double-replacement reactions generally occur between substances in aqueous solution. In order for a reaction to occur, one of the products is usually a solid precipitate, a gas, or a molecular compound such as water.

### Formation of a Precipitate

A precipitate forms in a double-replacement reaction when the cations from one of the reactants combine with the anions from the other reactant to form an insoluble ionic compound. When aqueous solutions of potassium iodide and lead (II) nitrate are mixed, the following reaction occurs.

$$2KI(aq)+Pb(NO_3)_2(aq) \rightarrow 2KNO_3(aq)+PbI_2(s)$$

There are very strong attractive forces that occur between $Pb^{2+}$ and $I^-$ ions and the result is a brilliant yellow precipitate. The other product of the reaction, potassium nitrate, remains soluble.

Formation of lead iodide precipitate.

### Formation of a Gas

Some double-replacement reactions produce a gaseous product which then bubbles out of the solution and escapes into the air. When solutions of sodium sulfide and hydrochloric acid are mixed, the products of the reaction are aqueous sodium chloride and hydrogen sulfide gas.

$$Na_2S(aq)+2HCl(aq) \rightarrow 2NaCl(aq)+H_2S(g)$$

### Formation of a Molecular Compound

Another kind of double-replacement reaction is one that produces a molecular compound as one of its products. Many examples in this category are reactions that produce water. When aqueous hydrochloric acid is reacted with aqueous sodium hydroxide, the products are aqueous sodium chloride and water.

$$HCl(aq) + NaOH(aq) \rightarrow NaCl(aq) + H_2O(l)$$

## Combustion Reactions

A combustion reaction is a major class of chemical reactions, commonly referred to as "burning." Combustion usually occurs when a hydrocarbon reacts with oxygen to produce carbon dioxide and water. In the more general sense, combustion involves a reaction between any combustible material and an oxidizer to form an oxidized product. Combustion is an exothermic reaction, so it releases heat, but sometimes the reaction proceeds so slowly that a temperature change is not noticeable. Good signs that you are dealing with a combustion reaction include the presence of oxygen as a reactant and carbon dioxide, water, and heat as products. Inorganic combustion reactions might not form all of the products but are recognizable by the reaction of oxygen.

Combustion doesn't always result in fire, but when it does, a flame is a characteristic indicator of the reaction. While the activation energy must be overcome to initiate combustion (i.e.: using a lit match to light a fire), the heat from a flame may provide enough energy to make the reaction self-sustaining.

### General Form of a Combustion Reaction

hydrocarbon + oxygen → carbon dioxide + water

### Examples of Combustion Reactions

Here are several examples of balanced equations for combustion reactions. It's important to remember that combustion reactions are easy to recognize because the products always contain carbon dioxide and water. In these examples, oxygen gas is present as a reactant, but trickier examples of the reaction exist where the oxygen comes from another reactant:

- Combustion of methane

  $CH_4(g) + 2\ O_2(g) \rightarrow CO_2(g) + 2\ H_2O(g)$

- Burning of naphthalene

  $C_{10}H_8 + 12\ O_2 \rightarrow 10\ CO_2 + 4\ H_2O$

- Combustion of ethane

  $2\ C_2H_6 + 7\ O_2 \rightarrow 4\ CO_2 + 6\ H_2O$

- Combustion of butane (commonly found in lighters)

  $2C_4H_{10}(g) + 13O_2(g) \rightarrow 8CO_2(g) + 10H_2O(g)$
- Combustion of methanol (also known as wood alcohol)

  $2CH_3OH(g) + 3O_2(g) \rightarrow 2CO_2(g) + 4H_2O(g)$
- Combustion of propane (used in gas grills and fireplaces)

  $2C_3H_8(g) + 7O_2(g) \rightarrow 6CO_2(g) + 8H_2O(g)$

## Complete versus Incomplete Combustion

Combustion, like all chemical reactions, does not always proceed with 100% efficiency. It is prone to limiting reactants the same as other processes. So, there are two types of combustion you are likely to encounter:

- Complete Combustion - Also called "clean combustion", clean combustion is oxidation of a hydrocarbon that produces only carbon dioxide and water. An example of clean combustion is burning of candle wax, where the heat from the wick vaporizes wax (a hydrocarbon), which reacts with oxygen in the air to release carbon dioxide and water. Ideally, all the wax burns so nothing remains once the candle is consumed. The water vapor and carbon dioxide dissipate into air.
- Incomplete Combustion - Also called "dirty combustion", incomplete combustion is hydrocarbon oxidation that produces carbon monoxide and/or carbon (soot) in addition to carbon dioxide. An example of incomplete combustion would be burning coal, where a lot of soot and carbon monoxide is released. Many of the fossil fuels burn incompletely, releasing waste products.

## References


- Decomposition-reaction, chemistry: byjus.com, Retrieved 24 June, 2019
- Barke, Hans-Dieter; Hazari, Al; Yitbarek, Sileshi (2008). Misconceptions in chemistry addressing perceptions in chemical education (Online-Ausg. ed.). Berlin: Springer. pp. 227–228. ISBN 3540709894.
- Combustion-reactions-604030: thoughtco.com, Retrieved 11 April, 2019
- Introductory-Chemistry, A-Double-Replacement-Reactions, A-Introductory-Chemistry, A-Chemical-Reactions: chem.libretexts.org, Retrieved 30 March, 2019

# 6
# Organometallic Chemistry

The branch of chemistry which studies the chemical compounds which contain bonds between carbon and metal atoms is known as organometallic chemistry. Most organometallic compounds are used as catalysts in a number of industries. The diverse applications of organometallic compounds have been thoroughly discussed in this chapter.

Organometallic chemistry is the study of organometallic compounds which contains metal-carbon bonds. Since many compounds without such bonds are chemically similar, an alternative may be compounds containing metal-element bonds of a largely covalent character. Organometallic chemistry combines aspects of inorganic chemistry and organic chemistry.

## Organometallic Compounds

Organometallic compound refers to any member of a class of substances containing at least one metal-to-carbon bond in which the carbon is part of an organic group. Organometallic compounds constitute a very large group of substances that have played a major role in the development of the science of chemistry. They are used to a large extent as catalysts (substances that increase the rate of reactions without themselves being consumed) and as intermediates in the laboratory and in industry. The class includes such compounds as ferrocene, a remarkably stable compound in which an iron atom is sandwiched between two hydrocarbon rings.

Organometallic compounds are typically discussed in terms of the metal as either main-group compounds or transition metal compounds. The main-group metals of organometallic compounds are typically considered to be those of the S-block (groups 1 and 2) and the heavier elements of the p-block (groups 13–15) in the periodic table of elements. The transition metals include those elements in the d- and f-blocks (groups 3–12).

The physical and chemical properties of organometallic compounds vary greatly. Most are solids, particularly those whose hydrocarbon groups are ring-shaped or aromatic, but some are liquids and some are gases. Their heat and oxidation stability vary widely. Some are very stable, but a number of compounds of electropositive elements such as lithium, sodium, and aluminum are spontaneously flammable. Many organometallic compounds are highly toxic, especially those that are volatile.

The properties of the organometallic compounds depend in large measure on the type of carbon-metal bonds involved. Some are ordinary covalent bonds, in which pairs of electrons are shared between atoms. Others are multicentre covalent bonds, in which the bonding involves more than two atoms. A third type are ionic bonds, in which the bonding electron pair is donated by only one atom. In donor-acceptor bonds, the metal atom is connected to hydrocarbons with multiple bonds between carbon atoms.

Where metal atoms form covalent bonds with carbon atoms, the electrons are usually shared unequally. As a result, the bond is polarized—one end is more negative than the other. The extent of polarization depends on the strength with which the metal atom binds electrons. Organometallic compounds range in polar power from methylpotassium, in which the bond is almost like certain ionic bonds, to lead, which bonds with carbon with very little polarization.

## Importance of Organometallic Compounds

Because of bond polarity, many organometallic compounds have reactivities that have made them important in chemical synthesis. The organomagnesium halides (Grignard reagents), for example, are used widely in synthetic organic chemistry, as are organolithium and organoboron compounds. Alkylaluminum compounds are also employed in organic synthesis. Used with titanium salts, they are important catalysts in the polymerization of unsaturated hydrocarbons, such as ethylene and propylene. The mechanism of action of the titanium-aluminum alkyl catalysts probably involves interaction between the titanium atoms and the double bonds of the hydrocarbons.

Organometallic compounds containing lead, tin, and mercury are all commercially significant. A large number of organotin compounds, for example, are used as pharmaceuticals, pesticides, stabilizers for polyvinyl chloride, and fire retardants. Methylmercury has caused severe pollution problems as a result of its toxicity. This fact has led to stringent controls on the discharge of mercury from chemical plants into rivers, lakes, and oceans.

Carbon monoxide reacts readily with many transition-metal atoms to form metal carbonyls, themselves a class of organometallics. One of the earliest to be discovered was tetracarbonylnickel, a volatile nickel compound that became the basis of a process for purifying nickel. Metal carbonyls are employed as catalysts in many reactions in the petrochemical industry.

## Defining Characteristics

A compound is regarded as organometallic if it contains at least one metal-carbon (M–C) bond where the carbon is part of an organic group. Typically, an organic group contains carbon-hydrogen (C–H) bonds; for example, the simple methyl group, $CH_3$, and larger homologs such as the ethyl group, $C_2H_5$, which attach to a metal atom through only one carbon atom. (Simple alkyl groups such as these are often abbreviated by the symbol R.) More elaborate organic groups include the cyclopentadienyl group, $C_5H_5$, in which all five carbon atoms can form bonds with the metal atom. The term metallic is interpreted broadly in this context; thus, when organic groups are attached to the metalloids such as boron (B), silicon (Si), germanium (Ge), and arsenic (As), the resulting compounds are considered to be organometallic along with those containing true metals such as lithium (Li), magnesium (Mg), aluminum (Al), and iron (Fe). The "metal" in an organometallic compound can include most elements, with the exception of nitrogen (N) and phosphorus (P) in group 15 and all the elements in groups 16 (the oxygen group), 17 (halogens), and 18 (noble gases).

| Common organic groups (ligands) in organometallic chemistry | | | |
|---|---|---|---|
| **Available electrons** | **Hapticity** | **Ligand** | **Metal-ligand Structure** |
| 1 | $\eta^1$ | Methyl, alkyl<br>$CH_3$,$CH_2R$ | 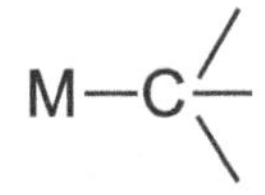 |
| 2 | $\eta^1$ | Alkylidene<br>(carbene) | 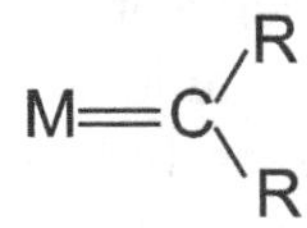 |
| 2 | $\eta^2$ | Alkene<br>$H_2C = CH_2$ | 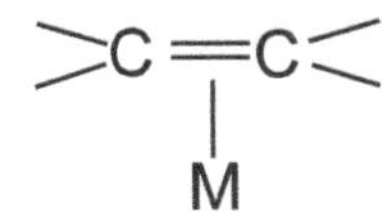 |
| 3 | $\eta^3$ | $\pi$ – allyl<br>$C_3H_5$ | 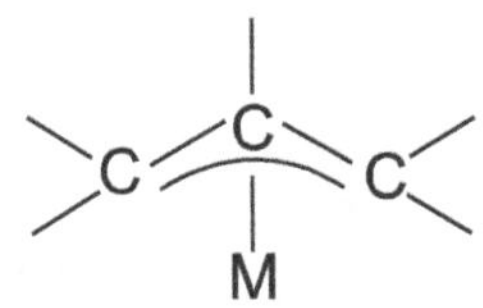 |
| 3 | $\eta^1$ | Alkylidyne<br>(carbyne)<br>C–R | $M\equiv C-R$ |
| 4 | $\eta^4$ | 1,3- butadiene<br>$C_4H_6$ | 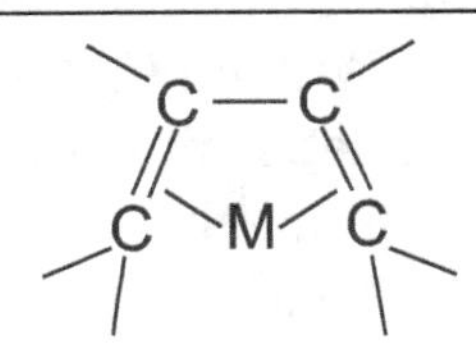 |
| 4 | $\eta^4$ | Cyclobutadiene<br>$C_4H_4$ | 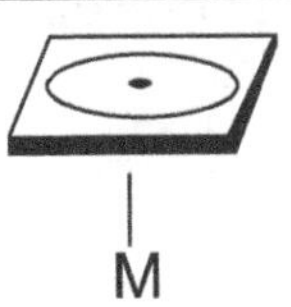 |
| 5<br>(3)<br>(1) | $\eta^5$<br>$\eta^3$<br>$\eta^1$ | Cyclopentadieyl<br>$C_5H_5$(Cp) | 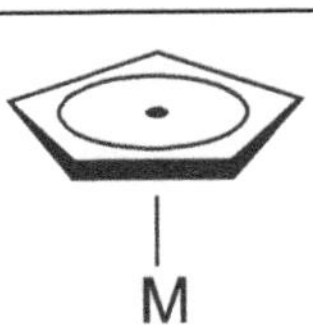 |
| 6 | $\eta^6$ | Benzene<br>$C_6H_6$ | 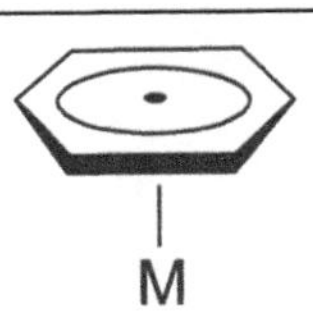 |
| 8<br>(6)<br>(4) | $\eta^8$<br>$\eta^6$<br>$\eta^4$ | Cyclooctatetraene<br>$C_8H_8$(cot) | 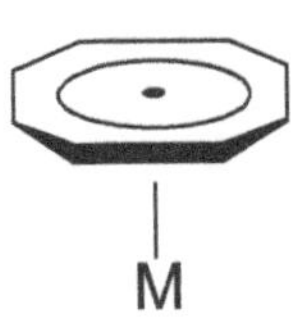 |

One example of an organometallic compound is trimethylboron, $B(CH_3)_3$, which contains three B–C bonds.

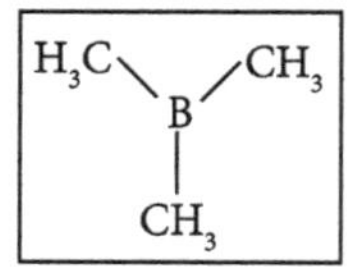

Another is ferrocene, $Fe(C_5H_5)_2$, which has a more elaborate structure with the iron atom sandwiched between two $C_5H_5$ rings. Some compounds with metal-carbon bonds are not regarded as organometallic, because the constituent carbon atom is not part of an organic group; two examples are metal carbides—such as $Fe_3C$, a hard solid that is a component of cast iron—and metal cyanide compounds—such as the deep-blue paint pigment Prussian blue, $KFe_2(CN)_6$.

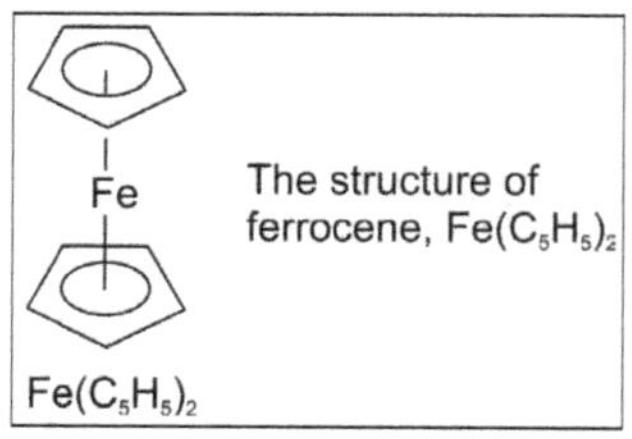


Ferrocene, a derivative of iron, is known as a sandwich compound because the iron atom is contained between two organic ring systems.

## Historical Developments

The first synthetic organometallic compound, $K[PtCl_3(C_2H_4)]$, was prepared by the Danish pharmacist William C. Zeise in 1827 and is often referred to as Zeise's salt. At that time, Zeise had no way of determining the structure of his new compound, but today it is known that the structure contains an ethylene molecule ($H_2C{=}CH_2$) attached through both carbon atoms to the central platinum (Pt) atom. The platinum atom also is bonded to three chlorine (Cl) atoms. The potassium ion, $K^+$, is present to balance the charge.

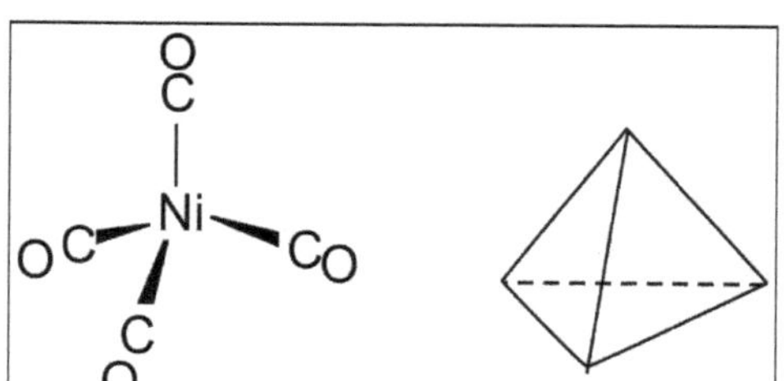

Tetracarbonylnickel, a type of metal carbonyl compound, has a high volatility and is extremely toxic.

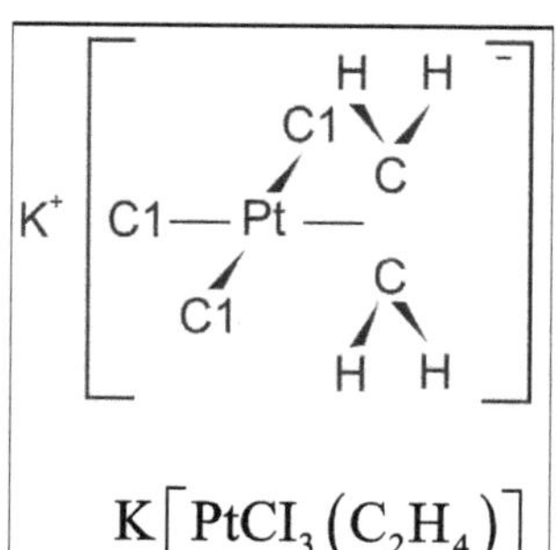

The attachment of the ethylene carbon atoms to the central platinum atom qualifies Zeise's salt as an organometallic compound. A development with a more immediate impact on the field of chemistry was the discovery in 1849 by the German-trained British chemist Edward C. Frankland of diethylzinc, $H_5C_2-Zn-C_2H_5$, which he showed is very useful in organic synthesis. Since then, an ever-increasing variety of organometallic compounds have been utilized in organic synthesis in both the laboratory and industry.

Another milestone in the development of the field was the discovery of tetracarbonylnickel by the German-educated British industrial chemist Ludwig Mond and his assistants in 1890. In 1951, German theoretical chemist Ernst Otto Fischer and British chemist Sir Geoffrey Wilkinson independently discovered the sandwich structure of the compound ferrocene. Their parallel discoveries led to the subsequent unveiling of other compounds with sandwich structures, and in 1973, Fischer and Wilkinson were jointly awarded the Nobel Prize for Chemistry for their contributions to the study of organometallic compounds. Since the 1950s, organometallic chemistry has become a very active field, marked by the discovery of new organometallic compounds along with their detailed structural and chemical characterization and their application as synthetic intermediates and catalysts in industrial processes. Two organometallics encountered in nature are the vitamin $B_{12}$ coenzyme, which contains a cobalt-carbon (Co−C) bond, and dimethylmercury, $H_3C-Hg-CH_3$, which is produced by bacteria to eliminate the toxic metal mercury. However, organometallic compounds are generally unusual in biological processes.

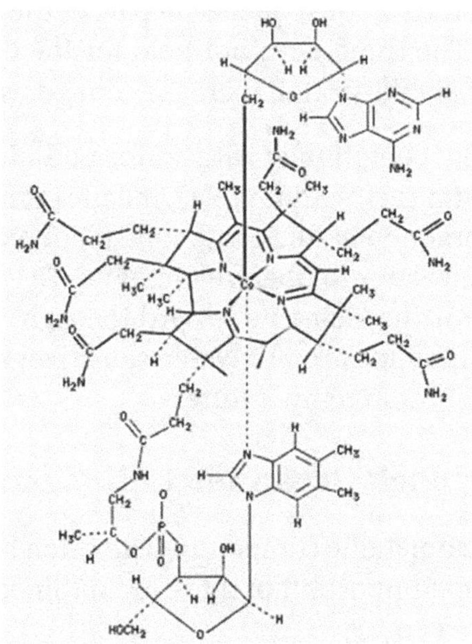

A schematic structure for vitamin $B_{12}$ coenzyme, which contains five nitrogen-cobalt bonds and one cobalt-carbon bond.

## S- and P-block Organometallic Compounds

The metal in main-group organometallic compounds can be any of the elements in the s block (i.e., groups 1 and 2) or any of the heavier elements in groups 13 through 15. (Groups 13–18 constitute the p block.) The elements at the borderline between the d block and p block—namely, zinc, cadmium, and mercury—will be discussed along with the p-block organometallics because of the similarity of their organometallic chemistry. In an internationally sanctioned system of nomenclature, the organic group is named first, followed by the metal, as in dimethylmercury. In writing the formula, this order is reversed, $Hg(CH_3)_2$. The organic groups, which are also called ligands, are named in the same way as for any organic compound. The number of carbon atoms on a group that are attached to the metal is indicated by the superscript in ηn. This convention is known as hapto nomenclature. A single point of attachment, $\eta^1$, is usually not explicitly indicated, as in the above formula for dimethylmercury, a monohapto species. The compound with the common name ferrocene has the systematic name bis($\eta^5$-cyclopentadienyl)iron, where the number of cyclopentadienyl ligands (two) is indicated by the prefix bis and the number of sites of attachment (five) for each of these is indicated by $\eta^5$. Ferrocene is thus called a pentahapto compound. The number of sites of attachment are also indicated in the formula $Fe(\eta^5\text{-}C_5H_5)_2$.

## The Stability and Reactivity of Organometallic Compounds

The stability and reactivity of organometallic compounds are associated with the nature of the organic ligands and the metal to which they are attached. In each of the main groups of the periodic table (groups 1, 2, and 13–15), the thermal stability of a given type of organometallic compound generally decreases from the lightest to the heaviest element in a group. For example, in compounds containing group-1 metals, methyllithium ($LiCH_3$) is much more stable than methylpotassium ($KCH_3$), and, in those with group-14 metals, tetramethylsilicon, $Si(CH_3)_4$, is stable in the absence of air at 500 °C (932 °F), whereas tetramethyllead, $Pb(CH_3)_4$, rapidly decomposes at that temperature. This trend in stability is a consequence in part of the decrease in M–C bond strength on going down within a group. The trend does not hold for the d-block elements (groups 3–12), where M–C bond strengths and stability often increase going down a group.

The reactivities of organometallic compounds with water and air vary widely. The highly active main-group metals such as lithium (Li), sodium (Na), magnesium (Mg), and aluminum (Al) form highly air- and water-sensitive organometallic compounds. For example, $A_{12}(CH_3)_6$ undergoes immediate and violent reaction with water to liberate methane ($CH_4$) gas, and it bursts immediately into flame on contact with air. For the elements toward the right of the periodic table (groups 14 and 15), the organometallic compounds are not water-sensitive; tetramethylsilicon, for example, does not react with water or air at room temperature.

## The Synthesis of s- and p-block Organometallic Compounds

Synthesis of s- and p-block organometallic compounds can often be accomplished by one of several general reaction types. The most important of these are outlined below.

## Formation of Alkyllithium and Grignard Reagents

The highly active metals combine with a halogen-substituted hydrocarbon to produce simple organometallic compounds. For example, methyllithium, an important reagent in organic synthesis, is produced commercially by following the reaction:

$$2Li + CH_3Cl \rightarrow LiCH_3 + LiCl$$

With other active metals, such as magnesium, aluminum, and zinc, the reaction generally yields the organometallic halide. A common reaction of this type is the synthesis of a Grignard reagent, an alkylmagnesium halide that finds wide use in organic synthesis (the s indicates that the metal is in solid form).

$$Mg(s)+CH_3\,Br \xrightarrow{(C_2H_5)2^O} CH_3MgBr$$

## Double Displacement

The synthesis of organometallic compounds by double displacement involves organometallic (MR) and binary halide (EX, where E is a metal or metalloid and X is a halogen) starting materials. It provides a convenient synthetic procedure that is widely used in the laboratory and to a lesser extent on a commercial scale. As the following examples illustrate, the organic group on the more active metal is transferred to the less active metal or metalloid. In this context the most common highly active metals are lithium, aluminum, and magnesium.

$$4Li(CH_3) + SiCl_4 \rightarrow 4LiCl + Si(CH_3)_4$$

$$Al_2(CH_3)_6 + 2BF_3 \rightarrow 2AlF_3 + 2B(CH_3)_3$$

## Redistribution

Double displacements involving the same central element are often referred to as redistribution reactions. A commercially important example is the redistribution of silicon tetrachloride and tetramethylsilicon (also known as tetramethylsilane) at elevated temperatures.

$$SiCl_4 + (CH_3)_4Si \rightarrow CH_3SiCl + (CH_3)_2SiCl_2 + (CH_3)_3SiH + ...$$

The products from this reaction can be separated by distillation. This reaction is performed industrially where $(CH_3)_2SiCl_2$ is removed from the equilibrating mixture and then hydrolyzed to produce the intermediates for silicone polymers, which have the form $-(Si(CH_3)_2-O)-n$.

## Hydrometallation

The addition of a metal hydride to a multiple bond is called hydrometallation, and it leads to the formation of a metal-carbon bond.

$$M-H + H_2C=CH_2 \rightarrow MH_2C-CH_3$$

This reaction is driven mainly by the high C–H bond strength relative to most E–H bond strengths. Two important hydrometallation reactions are hydroboration and hydrosilation, illustrated, respectively, by the following examples.

$$H_3BOR_2 + (H_3C)_2\,C=CH_2 \xrightarrow{R_2O} (H_3C)_2\,CH\text{-}CH_2\,BH_2$$

$$SiH_4 + (CH_3)_2\,C=CH_2 \xrightarrow{H_2PtCl_6} (CH_3)_2\,CH-CH_2\,SiH_3$$

In the hydroboration and hydrosilation of an unsymmetrical alkene, the boron or silicon binds to the carbon atom that has less-bulky substituents, and the smaller hydrogen atom binds to the carbon atom that has bulky substituents—$(CH_3)_2C$ in the above equations. Hydroboration was discovered and developed in the United States by Herbert C. Brown, who shared the Nobel Prize for

Chemistry in 1979 for this research. Both hydroboration and hydrosilation are widely used in the synthesis of complex organic molecules. In these applications, the B–C or Si–C bond is generally cleaved in a subsequent step to produce a product that is free of boron or silicon.

## Reduction

All organometallic compounds are potential reducing agents, and those of the electropositive elements are very strong reducing agents because the metal gives up electrons to the carbon, resulting in a polar M–C bond with a partial positive charge on the metal and a negative charge on the carbon. Organometallic compounds of highly electropositive elements such as lithium, sodium, and aluminum ignite spontaneously and sometimes explode on contact with air or other oxidizing agents. The useful organometallic reagents $Li(CH_3)$, $Zn(CH_3)_2$, $B(CH_3)_3$, and $Al_2(CH_3)_6$ are spontaneously flammable in air (pyrophoric). Accordingly, techniques have been developed to handle these and other pyrophoric compounds. Glass reaction vessels sealed from the atmosphere and purged with nitrogen gas are commonly used for handling air-sensitive organometallic compounds in the laboratory. Large quantities of pyrophoric compounds such as $Al_2(C_2H_5)_6$ are routinely handled with ease in the chemical industry by using closed metal reactors for the production of these and other much less reactive compounds. Organometallic compounds have reduced reactivity when the metallic component is not highly electropositive and when the metal is completely surrounded by attached groups. For example, elevated temperatures are required to initiate combustion with $Si(CH_3)_4$ and $Sn(CH_3)_4$, and at room temperature they can be handled in air.

## Carbanion Character

The partial negative charge of an organic group bonded to a highly active metal results in a distinctive pattern of reactivity that is frequently referred to as nucleophilic or carbanion character. Thus, organometallic compounds containing highly active (electropositive) metals, such as lithium, magnesium, aluminum, and zinc, react rapidly and completely with water, liberating a hydrocarbon in the process. For example, dimethylzinc liberates methane gas along with solid zinc hydroxide.

$$Zn(CH_3)_2 + 2H_2O \rightarrow Zn(OH)_2 + 2CH_4$$

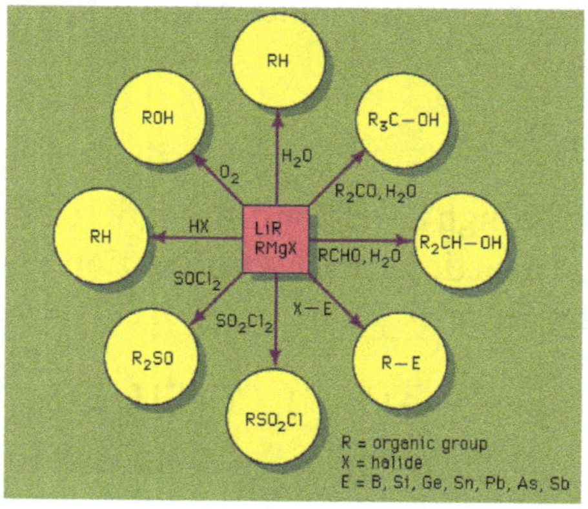


Highly reactive organometallic reagents, such as alkyllithium and Grignard reagents (LiR and RMgX, respectively), have multiple synthetic applications.

The above hydrolysis of dimethylzinc can be viewed as a transfer of a slightly acidic $H^+$ from water to the strongly basic carbanion $CH_3^-$ in dimethylzinc.

Alkyllithium, alkylaluminum, and alkylmagnesium compounds are the most common carbanion reagents in laboratory-scale synthetic chemistry; carbanion character is greatly diminished for the less metallic elements boron and silicon. The nucleophilic character of organometallic compounds of active metals has many synthetic applications. For example, the organic group in organometallic compounds of active metals attacks the carbonyl carbon of a ketone, and upon hydrolysis a tertiary alcohol results. Similarly, aldehydes can be converted to secondary alcohols by reaction with an organometallic reagent followed by hydrolysis. Double displacement reactions can be used to prepare sulfones ($R_2SO_2$) and sulfoxides ($R_2SO$) by treating thionyl chloride ($SOCl_2$) or sulfuryl dichloride ($SO_2Cl_2$) with an alkyllithium or a Grignard reagent.

One consequence of the carbanion character of organometallic compounds containing active metals is the protolysis (proton-transfer) reaction that takes place with very weak protonic acids, including water. Alcohols react in a manner similar to the reaction of water, and this provides a convenient way of introducing an alkoxide (OR) substituent into an organometallic compound.

$$(C_2H_5)_3Ga + HOCH_3 \rightarrow [(C_2H_5)_3GaOHCH_3] \rightarrow (C_2H_5)_2Ga(OCH_3) + C_2H_6$$

The rate of reaction decreases with bulky organic groups on the alcohol. For example, tert-butyl alcohol, reacts slowly with most active organometallics, and it is therefore employed in the laboratory to safely destroy reactive organometallic wastes.

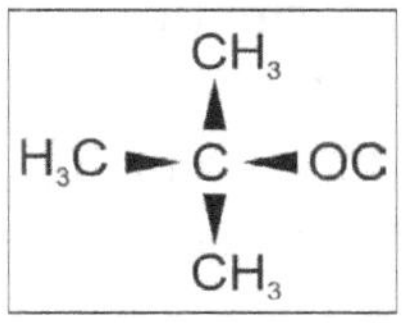


## β-hydrogen Elimination

This is a common organometallic reaction in which a hydrogen atom on a carbon atom that is one position removed from the metal (the β position) transfers to the metal with the liberation of an alkene. The following example shows the formation of ethylene, $C_2H_4$.

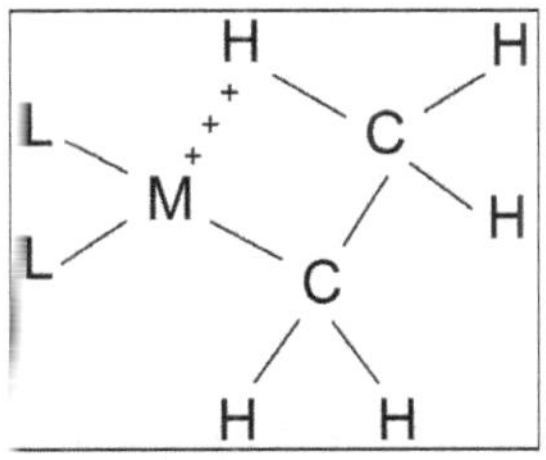


Compounds with an accessible central metal atom tend to readily undergo β-hydrogen elimination.

$$\underset{\alpha}{M\text{-}CH_2} - \underset{\beta}{CH_3} \rightarrow M\text{-}H + H_2C{=}CH_2$$

This reaction is the reverse of the addition of an M−H bond to an alkene and under some conditions significant equilibrium concentrations of both reactants and products are observed. The β-hydrogen elimination reaction is thought to proceed through the transition state, and as might be expected from this reaction mechanism, compounds with an accessible central metal atom tend to undergo β-hydrogen elimination readily. For example, it occurs with trialkylaluminum compounds but not with tetraalkylsilicon compounds, in which reaction of the silicon atom is hindered by the bulky substituents.

## D- and F-block Organometallic Compounds

The nomenclature for organometallic compounds of the d-block elements, also known as transition metals, follows rules similar to those described for the main-group organometallics. Groups attached to the metal (ligands) are named first, followed by the metal. When a ligand is attached to the metal by more than one carbon atom, the number of sites of attachment is indicated by n in $\eta^n$.

## Metal Carbonyls

Many of the early discoveries in d-block organometallic chemistry involved the metal carbonyls—i.e., compounds consisting of a metal atom bonded to one or more carbon monoxide (CO) ligands. Mond's discovery of the first simple metal carbonyl, tetracarbonylnickel, $Ni(CO)_4$, at the end of the 19th century was quickly followed by a series of discoveries in his laboratory and elsewhere showing that most of the d-block metals form neutral homoleptic carbonyls. (The term homoleptic refers to identical groups attached to a central atom.) The remarkable volatility of tetracarbonylnickel, whose boiling point is 43 °C, prompted one of Mond's contemporaries to state that "Mond put wings on metals." The ease of formation of tetracarbonylnickel (from metallic nickel and carbon monoxide) and its high volatility led Mond to develop an efficient industrial process for separating nickel and cobalt, which often occur together in their ores. This process requires diligent safety measures because tetracarbonylnickel is extremely toxic.

## The Structure of Metal Carbonyls

The carbonyl ligands in the tetracarbonylnickel molecule project toward the vertices of a tetrahedron, and thus the structure is referred to as tetrahedral. Similarly, the six carbonyl ligands in hexacarbonylchromium project toward the vertices of an octahedron.

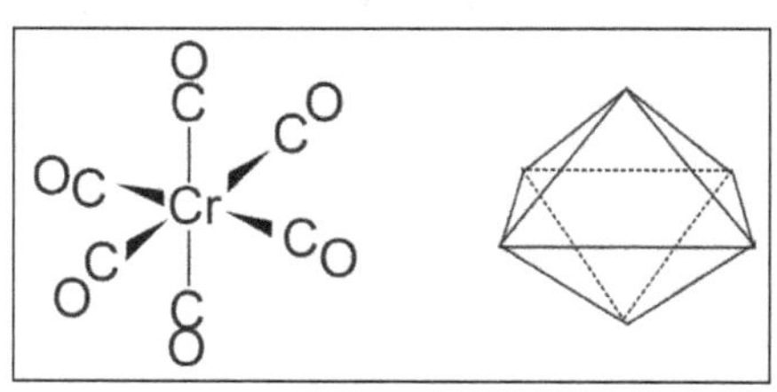

Many other metal carbonyls contain two or more metal atoms, such as decacarbonyldimanganese and octacarbonyldicobalt, shown here.

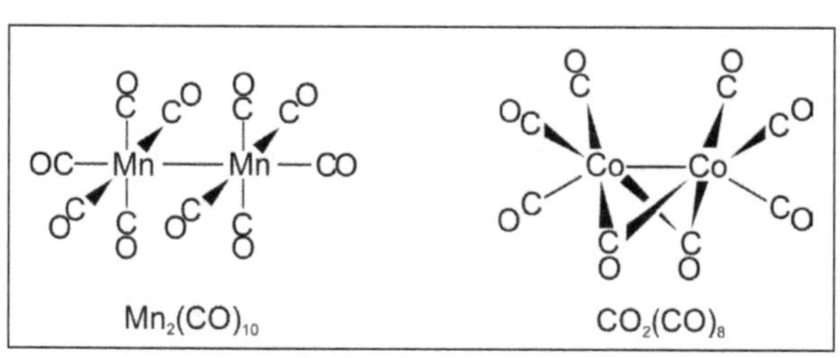


$Mn_2(CO)_{10}$ $CO_2(CO)8$

In all these structures carbon monoxide is connected to the metal through its carbon atom. When more than one metal atom is present, as in octacarbonyldicobalt, the carbon of the carbonyl ligand may bridge between metal atoms.

The formulas of most homoleptic metal carbonyl compounds conform to the rule that each metal atom in the metal carbonyl molecule must have 18 valence electrons. The valence electrons represent the outer electrons on the metal plus those from the ligand. For example, the electron count for $Fe(CO)_5$ includes eight from the iron atom (it is in group 8 and hence has eight valence electrons) and two from each carbonyl ligand, giving a total of 18 electrons. This 18-electron rule applies to many organometallic compounds of the d-block metals other than carbonyls, but there are exceptions in the organometallic chemistry of the d-block metals. The most notable exceptions are metals on the far left of the d block (e.g., $V(CO)_6$ contains 17 electrons) and on the right of the d block where the platinum metals—rhodium (Rh), iridium (Ir), palladium (Pd), and platinum (Pt)—often exhibit a 16-electron count in their organometallic compounds.

## Zero-oxidation-state Metal Carbonyls

The central metal in a neutral metal carbonyl, such as those described above, is assigned an oxidation state of zero, quite unlike the case in simple inorganic compounds in which positive oxidation states are the norm, as, for example, $Fe^{3+}$ in $FeCl_3$ or $Ni^{2+}$ in $NiBr_2$. Unlike the free metals, which also have a zero oxidation state, many carbonyls are soluble in a variety of simple organic solvents and are highly reactive. Because of these chemical and physical properties, the metal carbonyls are convenient starting materials for the synthesis of compounds with the metal atom in a zero or low oxidation state. One simple reaction is the substitution of other ligands such as triethylphosphine, $P(Et)_3$, for CO (Et is a common abbreviation for the ethyl group, $-C_2H_5$).

$$Cr(CO)_6 + P(Et)_3 \rightarrow Cr(CO)_5P(Et)_3 + CO$$

Because most other ligands are far bulkier than carbon monoxide and also because of differences in bonding properties, it is often not possible to replace all the carbonyl ligands by groups such as triethylphosphine, and a mixture of products with varying degrees of substitution can result.

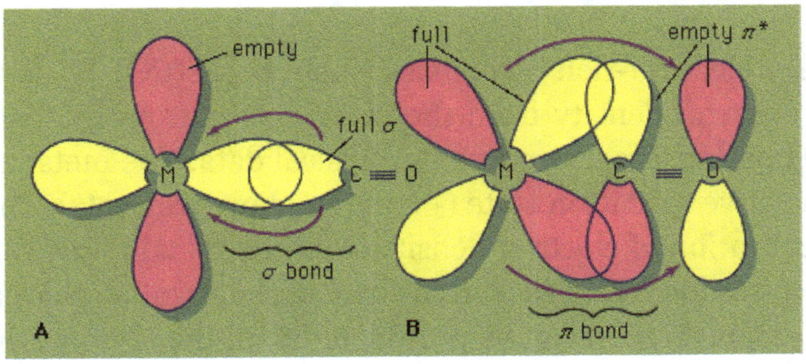


Bonding of the CO ligand to a metal atom(A) The σ bond formed by the partial donation of a pair of electrons from an orbital largely localized on the carbon atom to a d orbital on the metal atom. (B) Back π bonding from filled d orbitals on the metal atom into an empty π* orbital on the CO ligand. In most cases, the net back π bonding predominates, and electron density is transferred from the metal to the CO ligand.

The remarkable ability of the carbonyl ligand to stabilize compounds with metals in the zero oxidation state has led to detailed studies of M–CO bonds. A minor part of the M–CO bonding can be attributed to the tendency of the carbonyl ligand to donate a pair of electrons to the metal, which is the principal mode of interaction for most ligands with metal cations in classical coordination compounds such as $[Co(NH_3)_6]^{3+}$. The second mode of interaction with the metal is the simultaneous back-donation of electron density from the metal to the carbonyl ligand, which is called back π bonding.

## Metal Carbonyl Anions

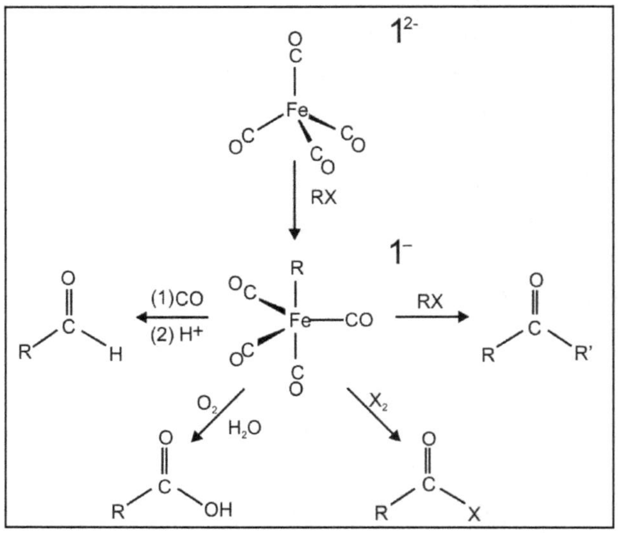


The organometallic reagent $Fe(CO)_4^{2-}$ can undergo multiple reactions in the synthesis of organic molecules.

More remarkable than the formation of zero-oxidation-state metal carbonyls is the reduction of many of these carbonyl compounds to metal carbonyl anions, in which the metal has a negative oxidation state. The following example demonstrates that the two-electron reduction by sodium metal is accompanied by the loss of the two-electron donor carbonyl ligand, and so the 18-electron count on iron is preserved; the solvent is tetrahydrofuran (THF).

$$2Na + Fe(CO)_5 \xrightarrow{THF} Na_2\left[Fe(CO)_4\right] + CO$$

The stabilization of very electron-rich complexes, such as $[Fe(CO)_4]^{2-}$, is attributed to the back π bonding that shifts electron density from the metal to the carbonyl ligands, and this view is substantiated by C–O bond distances and other physical data. The metal atom in these carbonyl anions is assigned a negative oxidation state (–2 in the above example). This formalism does not acknowledge the delocalization of electron density from the metal to the ligand, but the chemical properties of the carbonyl anions do suggest that some of the negative charge resides on the metal. For example, a metal carbonyl anion can be protonated with the $H^+$ ion, which generally attaches to the central metal and not a carbonyl ligand, as in the following example.

$$[Fe(CO)_4]^{2-} + HCl \rightarrow [HFe(CO)_4]^- + Cl^-$$

Owing to their high reactivity, carbonyl anions are useful starting materials for the synthesis of other organometallic compounds, and this accounts for their applications in organic synthesis. For example, $[Fe(CO)_4]^{2-}$ is used to extend a carbon chain by transfer of the carbonyl substituent, producing aldehydes, ketones, or carboxylic acids.

## Compounds with Metal-carbon Bonds

The variety of hydrocarbon ligands found in d-block organometallic chemistry range from simple σ-bonded alkyl ligands, double-bonded carbenes, and triply bonded carbynes to a host of polyene ligands.

## Simple Alkyl Ligands

A simple alkyl ligand forms an M–C single bond, and in doing so the alkyl group acts as a one-electron monohapto ligand. Many such compounds are known, but they are less common in the d block than in the s and p blocks. This may in part be a result of the modest M–C bond strengths. Another reason for the limited stability of alkyl ligands in many d-block complexes is a set of reactions that can be quite rapid, such as β-hydrogen elimination, CO insertion, and reductive elimination. As described below, these reactions transform simple hydrocarbon ligands into other groups.

The β-hydrogen elimination reaction is an important feature of the hydrocarbon chemistry of both d- and p-block organometallics. The reaction consists of the abstraction of a hydrogen atom from the organic ligand with the formation of two products, one of which contains a metal-hydrogen bond and the other of which is an alkene.

$LnM{-}CH_2CH_3 \rightarrow LnM{-}H + H_2C{=}CH_2$(Ln represents the n ligands not involved in the hydrogen elimination.) The reverse of this reaction, alkene insertion into the M–H bond, is illustrated by the hydroboration and hydrosilation reactions discussed above. Both the β-hydrogen elimination and addition of M–H across a C=C double bond are thought to proceed through a cyclic intermediate involving a three-centre, two-electron bond where a hydrogen atom bridges between the carbon and the metal atoms.

Since a compound with no hydrogen atoms on the β-carbon atom cannot undergo β-hydrogen elimination, benzyl, $CH_2(C_6H_5)$, and methyltrimethylsilyl, $CH_2Si(CH_3)_3$ ligands are generally more robust than ethyl ligands when attached to d-block metal atoms. Similarly, the lack of β-hydrogen atoms on the methyl group accounts for the greater stability of complexes containing the methyl ligand rather than the ethyl ligand.

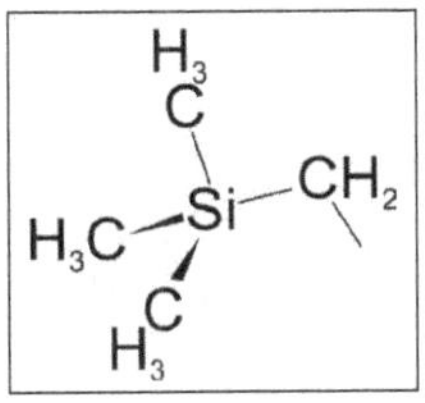

A reaction frequently referred to as CO insertion leads to carbon-carbon bond formation between the carbon atom of a carbonyl ligand and the carbon atom of an alkyl ligand, which is the methyl group in the following example.

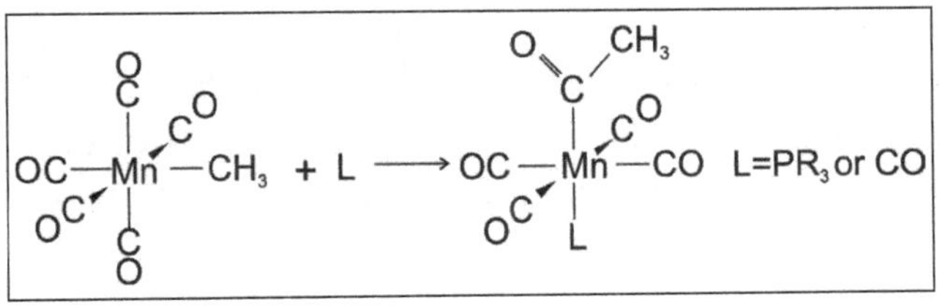

The CO insertion reaction is involved in all the transformations of $[Fe(CO)_4R]^-$ into organic molecules.

Another type of reaction that can transform an attached organic ligand (as well as other groups) is reductive elimination.

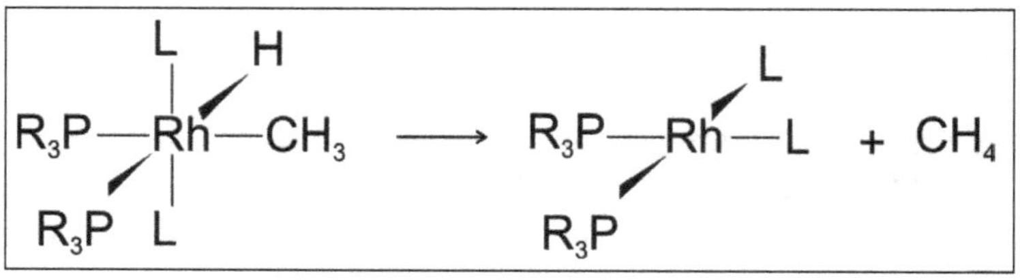

The converse of reductive elimination is oxidative addition.

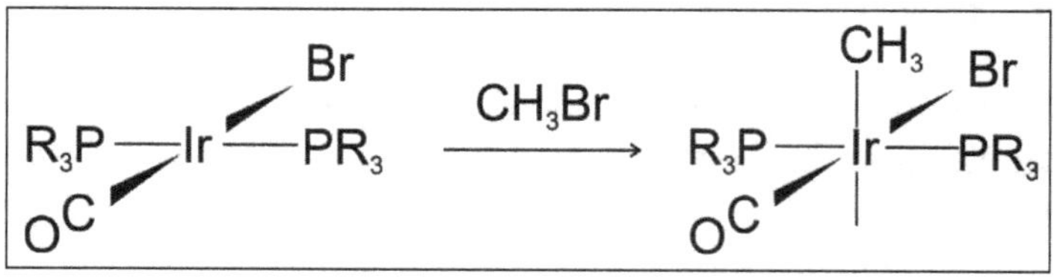

The reactions discussed above, insertion of C=C into an M−H bond, β-hydrogen elimination, and CO insertion, are often employed for the synthesis of organic molecules in the laboratory and in industry. They also account for some of the individual steps in some important catalytic cycles.

## Alkylidene Ligands

Alkylidene ligands, such as $CH_2$, CHR, or $CR_2$, form the M=C d-p double bonds (i.e., bonds between the d orbitals of the metal and the p orbitals of the carbon), and their metal compounds are often called carbenes. The first stable metal carbene complex was discovered in the laboratory of the German chemist Ernst O. Fischer in Munich by the reaction of a metal carbonyl with an alkyllithium compound. In this reaction, the alkyl group is transferred (as an $R^-$ group) to the carbon atom of the coordinated carbonyl, and subsequent addition of carbocation reagent results in attachment of an $R^+$ group to the oxygen.

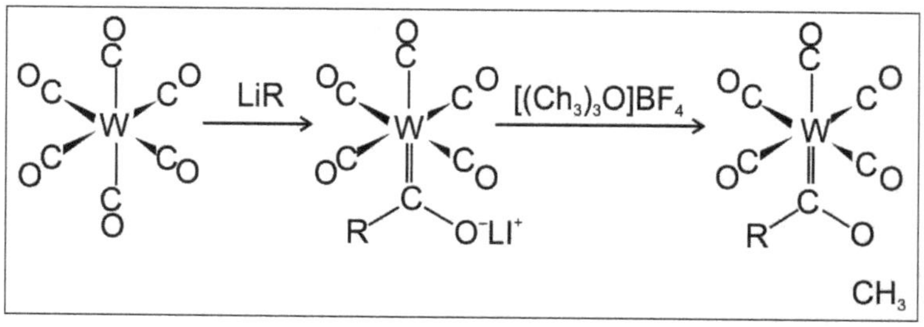

This type of carbene complex is common for the atoms of metals in groups 6–8, and they are called Fischer carbenes. The Fischer carbenes can be modified by electron-rich groups. For example, the attack of an amine on the electron-poor carbon atom of a Fischer carbene results in the displacement of the OR group to yield a new carbene (Me represents the methyl group, $-CH_3$).

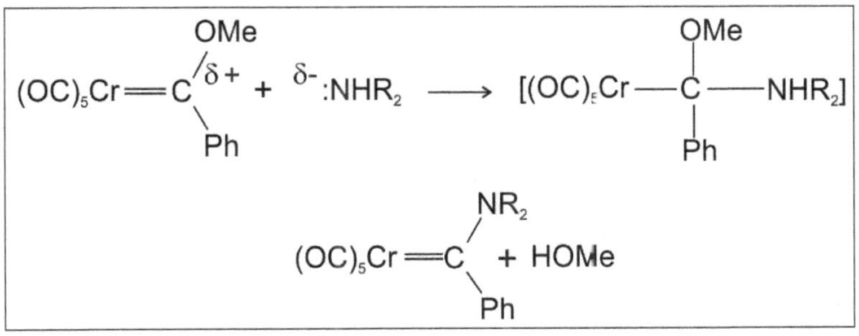

The attack of an electron-rich amine (indicated by δ- in the above equation) on the carbon atom in the Fischer carbene is attributed to the significant electronegativity of the middle to late d-block metals, which makes the carbon atom of the carbene electron-poor (indicated by δ+ in the above equation).

The reactions of Fischer carbene complexes with alkynes has considerable utility in organic synthesis. For example, naphthyl compounds (i.e., those derived from the fused ring system $C_{10}H_8$) can be synthesized by the reaction of methoxy phenyl Fischer carbenes with an alkyne.

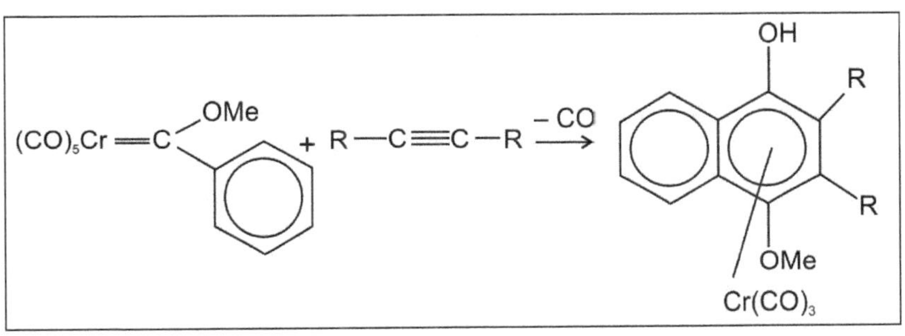

The chromium can be removed from the organic product by mild oxidation. Another route to a carbene is the deprotonation of an alkyl ligand, producing a carbene that contains only hydrocarbon ligands.

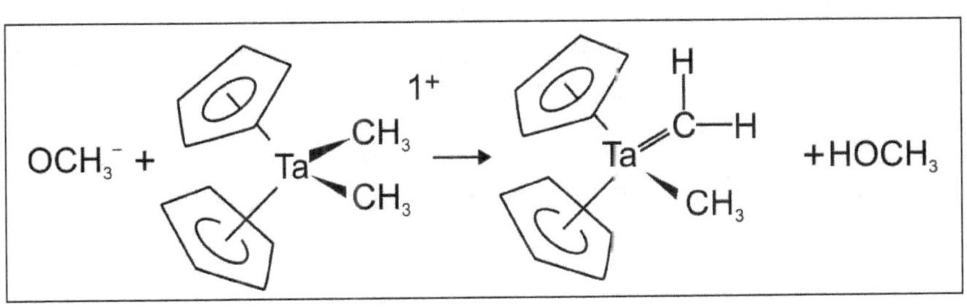

These ligands that contain only carbon and hydrogen are commonly attached to metal atoms from the early part of the d block such as titanium (Ti) and tantalum (Ta). The complexes are known as Schrock carbenes for their discoverer, American chemist Richard Schrock. The chemistry and spectroscopy of the Schrock carbenes indicate that these compounds have the opposite polarity of the Fischer carbenes. The carbon behaves as if it were electron-rich, because the $M^{\delta+}=C^{\delta-}$ bond is polarized so as to put negative charge on the metal-bound carbon, for the early d-block metals readily give up electrons. As a result, the carbon attached to the metal atom in a Schrock carbene reacts with electron-seeking reagents, such as $Me_3Si^{\delta+}Br^{\varepsilon+}$, at the carbene carbon (Cp is a common abbreviation for the cyclopentadienyl ligand, $-C_5H_5$).

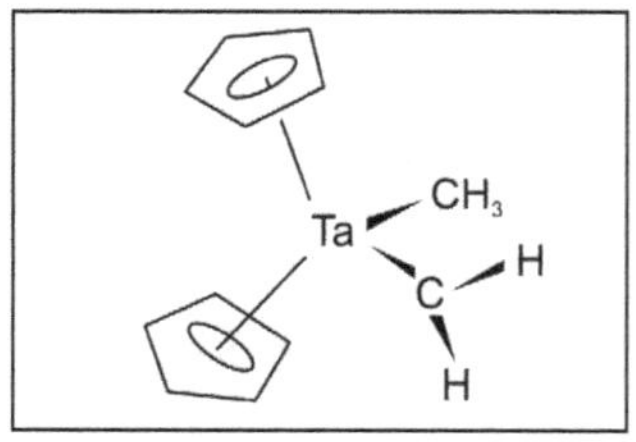


$$Cp_2Ta = (CH_2)(CH_3) + Me_3SiBr \rightarrow \left[Cp_2Ta(CH_2SiMe_3)(CH_3)\right]^+ + Br^-$$

An interesting reaction of the Schrock carbenes is the alkene metathesis reaction:

$$C_PCI_2Ta = C(H)CMe_3 + H_2C = CH_2 \rightarrow$$
$$C_PCI_2Ta = CH_2 + H_2C = C(H)(CMe_3)$$

This reaction appears to proceed through a four-membered ring intermediate containing carbon atoms and the metal atom.

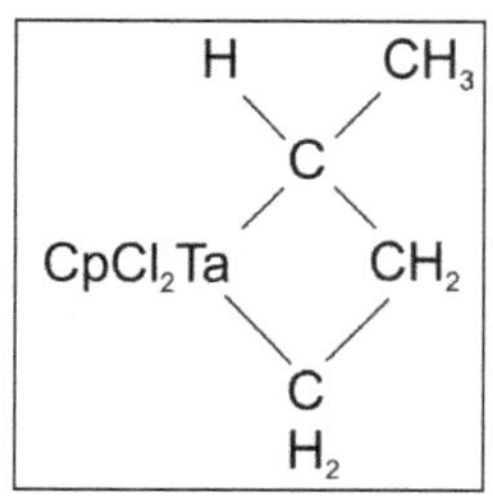


## Alkylidyne Ligands

Alkylidyne ligands have the general formula CH or CR. They are bound to the metal by an M≡C triple bond involving one σ bond and two d-p π bonds. The simplest member of this series is methylidyne, CH, and the next simplest, $CCH_3$, is ethylidyne. One route to methylidynes, discovered in Fischer's laboratory, involves the abstraction of an alkoxide group (OR) from a Fischer carbene by $BBr_3$.

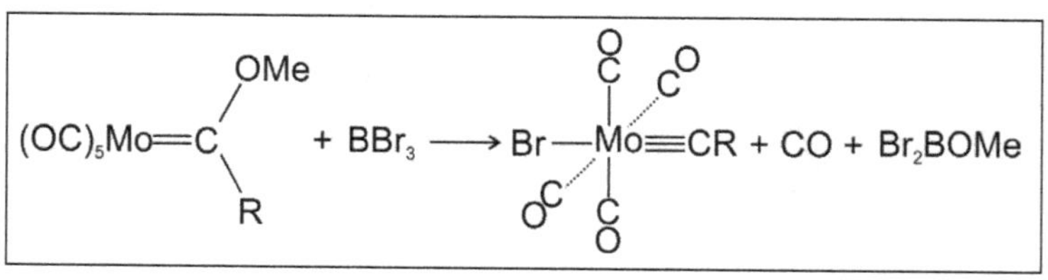


## Alkene and Alkyne Ligands

An alkene ligand contains a π bond between carbon atoms, C=C, which can serve as an electron pair donor in a metal complex, as in the case of Zeise's salt. This complex may be prepared by bubbling ethylene, $C_2H_4$, through an aqueous solution of $[PtCl_4]^{2-}$ in the presence of divalent tin, Sn(II), which aids in the removal of the chloride ion ($Cl^-$) from the coordination sphere of the divalent platinum, Pt(II).

$$K_2[PtC1_4] + H_2C{=}CH_2 \xrightarrow{SnC1_2} K[PtC1_3](\eta^2\text{-}C_2H_4)] + KC1$$

The alkene ligand bonds to the metal centre by both electron donation and acceptance, similar to the situation with carbon monoxide. Electron donor-and-acceptor character between the metal and the alkene ligand appear to be fairly evenly balanced in most ethylene complexes of the d metals.

The allyl ligand, $-CH_2-CH=CH_2$, can bind to a metal atom in either of two configurations: as an $\eta^1$-ligand or an $\eta^3$-ligand. Because of this versatility in bonding, $\eta^3$-allyl complexes are often highly reactive. Examples of $\eta^1$- and $\eta^3$-allyl complexes are, respectively, shown here.

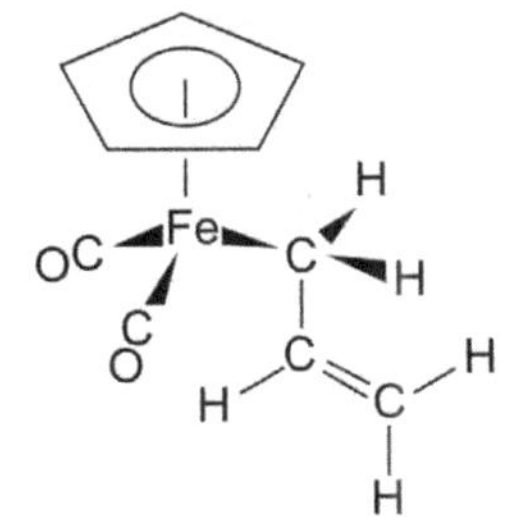


$Fe(\eta^1\text{-}C_3H_3)(CO)_2(\eta^5\text{-}Cp)$

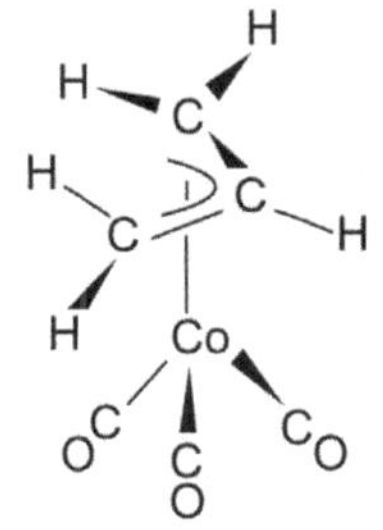


$Co(\eta^3\text{-}C_3H_5)(CO)_3$

Acetylene, H−C≡C−H, has two π bonds and hence is a potential four-electron donor. Substituted acetylenes form very stable polymetallic complexes in which the acetylene can be regarded as a four-electron donor. An example is $\eta^2$-diphenylethynehexacarbonyldicobalt, in which four of the six electrons in the triple bond of the ethyene ligand, R−C≡C−R, are shared with the two cobalt atoms (Ph represents the phenyl ligand, $-C_6H_5$). As in this example, the alkyl or aryl groups (R) on the acetylene impart stability to the metal complex—in contrast to simple acetylene (HC≡CH) complexes, where the hydrogen atoms are reactive.

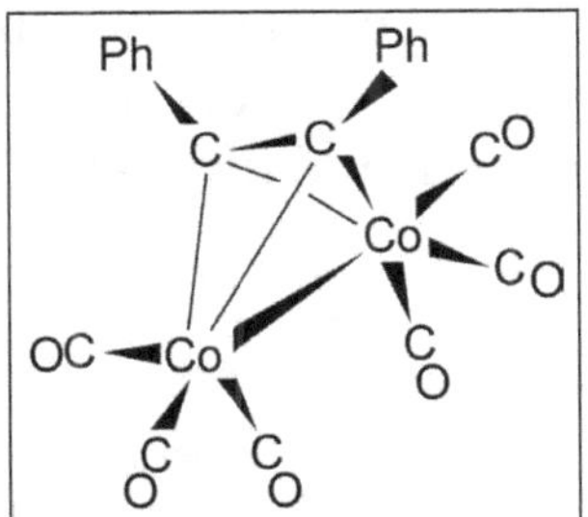


$Co_2(Phc_2Ph)(CO)_6$

## Polyene Ligands

Diene (−C=C−C=C−) and larger polyene ligands present the possibility of several points of attachment to a metal atom. The resulting polyene complexes are usually more stable than the equivalent monohapto complex with individual ligands. For example, bis($\eta^4$-cycloocta-1,5-diene)nickel is more stable than the corresponding complex containing four ethylene ligands.

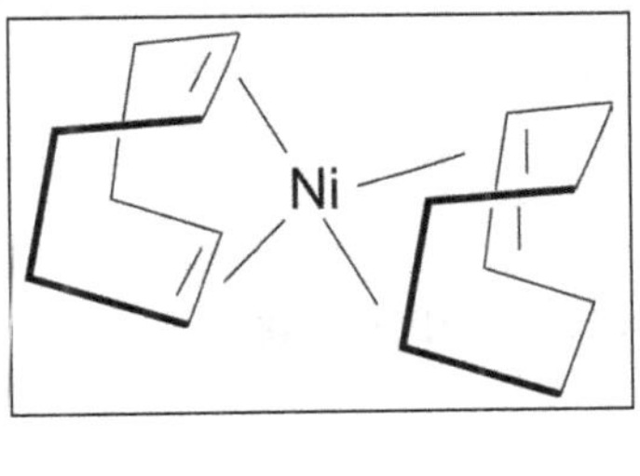

Cycloocta-1,5-diene (cod), a fairly common ligand in organometallic chemistry, is introduced into the metal coordination sphere by ligand displacement reactions; for example,

$$PdCl_2(NCPh)_2 + cod \rightarrow codPdCl_2 + 2NCPh.$$

Metal complexes of cod are often used as starting materials because the cod ligand can bind in various ways to the metal and the complexes are intermediate in stability. Many of them are sufficiently stable to be isolated and handled, but cod and similar ligands can be displaced by stronger ligands. For example, $Ni(cod)_2$ reacts with CO to form $Ni(CO)_4$ and the free cod molecule. This reaction is a convenient source of the extremely toxic $Ni(CO)_4$, for it can be generated directly in a flask where it is then available to undergo a subsequent reaction.

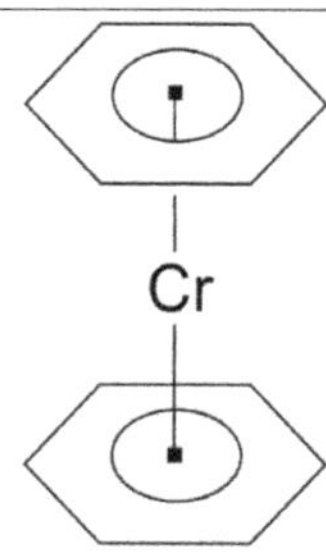


$Cr(\eta^6\text{-}C_6H_6)_2$

## Cyclic Polyene Ligands

These rings, which have alternating double and single bonds, are among the most important ligands in organometallic chemistry; the most common members of this group range from cyclobutadiene ($C_4H_4$) to cyclooctatetraene ($C_8H_8$). Their organometallic compounds include the metallocenes ferrocene and bisbenzenechromium and bis(cyclooctatrienyl)uranium (commonly called uranocene), shown here.

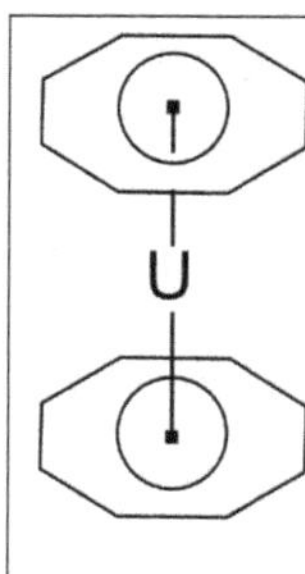


$U(\eta^8\text{-}C_8H_8)_2$

A metallocene consists of a metal atom between two planar polyhapto rings (as in ferrocene), and because of this structure they are informally called sandwich compounds. Cyclic polyenes are also known to form complexes in which they bind to a metal atom through some but not all of their carbon atoms.

The cyclobutadiene ligand is a four-electron donor. It is unstable as the free (i.e., uncombined) hydrocarbon, but it is known to exist in stable complexes, including $Ru(C_4H_4)(CO)_3$.

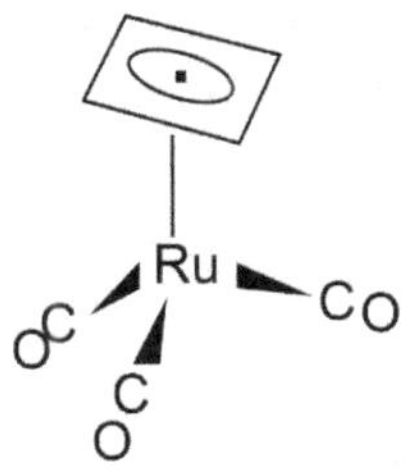

$Ru(C_4H_4)(CO)_3$

This is one of many cases in which coordination to a metal atom stabilizes an otherwise unstable molecule. Because of its instability, cyclobutadiene must be generated in the presence of the metal to which it is to be coordinated. This can be accomplished in several ways, one of which is the dimerization of a substituted acetylene. Interestingly, the $C_4R_4$ is bound to the cobalt in preference to the trimerization product, $C_6R_6$.

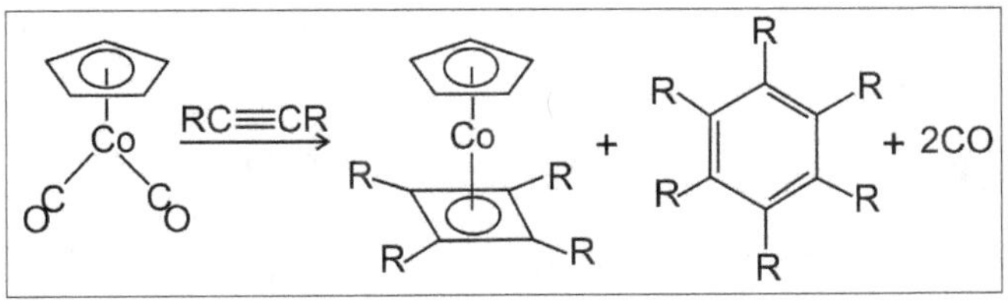

The cyclopentadienyl ligand ($C_5H_5$, abbreviated Cp) has played a major role in the development of organometallic chemistry. In some metal cyclopentadienyl compounds, the metal is bonded to only one of the five carbon atoms, and in these complexes the Cp is designated as a monohapto, $\eta^1$-, ligand, which contributes one electron to form a σ bond with the metal, as in:

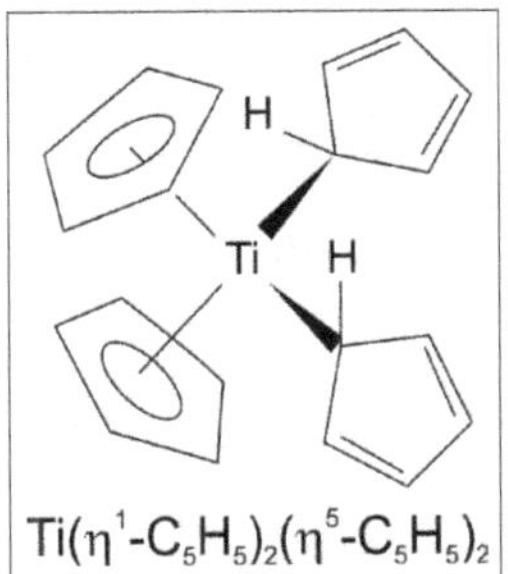

$Ti(\eta^1-C_5H_5)_2(\eta^5-C_5H_5)_2$

Others contain a trihapto ($\eta^3$-) cyclopentadienyl ligand, which donates three electrons. The most common case, however, is when Cp is a pentahapto ligand contributing five electrons. Two of the bonding modes for Cp are illustrated in the following structure, which contains both $\eta^3$- and $\eta^5$-$C_5H_5$ ligands.

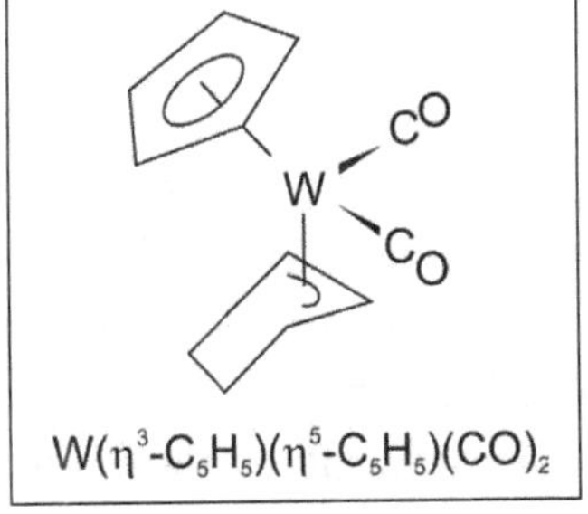

$W(\eta^3-C_5H_5)(\eta^5-C_5H_5)(CO)_2$

The bis($\eta^5$-cyclopentadienyl)-sandwich complexes of iron, cobalt, and nickel are readily prepared by the reaction of sodium cyclopentadienide with the corresponding d-metal halide.

$$2Na[C_5H_5] + MnCI_2 \xrightarrow{THF} Mn(C_5H_5)_2 + 2NaCI$$

Because of their great stabilities, the 18-electron group-8 compounds ferrocene, ruthenocene, and osmocene maintain their ligand-metal bonds under rather harsh conditions, and it is possible to carry out a variety of reactions on the cyclopentadienyl ligands while they are attached to the central metal. In some cases, they undergo reactions similar to those of simple aromatic hydrocarbons, such as Friedel-Crafts substitution, which is a characteristic reaction of benzene, $C_6H_6$.

$$CIC\overset{O}{\overset{\|}{-}}CH_3 + Fe(\eta^5 - C_5H_5) \xrightarrow{AICI_3} + Fe(\eta^5 - C_5H_5)\left(\eta^5 - C_5H_4\overset{O}{\overset{\|}{C}}CH_3\right) + HCI$$

It is also possible to replace the hydrogen atom on a $C_5H_5$ ring with a lithium atom using the highly reactive reagent butyllithium.

$$LiC_4H_9 + Fe(\eta^5\text{-}C_5H_5)_2 \rightarrow Fe(\eta^5\text{-}C_5H_5)(\eta^5\text{-}C_5H_4Li) + C_4H_{10}$$

This lithiated product is an excellent starting material for the synthesis of other ring-substituted products.

A closely related set of so-called bent sandwich compounds, in which the Cp rings are not parallel, are important in the organometallic chemistry of the early and middle d-block elements and the f-block elements (lanthanoids and actinoids). The Schrock carbene $Ta(\eta^5\text{-}C_5H_5)_2(CH_3)(CH_2)$, shown above, is one such example. Bent sandwich compounds are important in the organometallic chemistry of the f-block elements, but to achieve stability the pentamethylcyclopentadienyl ligand, $C_5(CH_3)_5$, is generally employed with these elements, as, for example, in the following uranium compound.

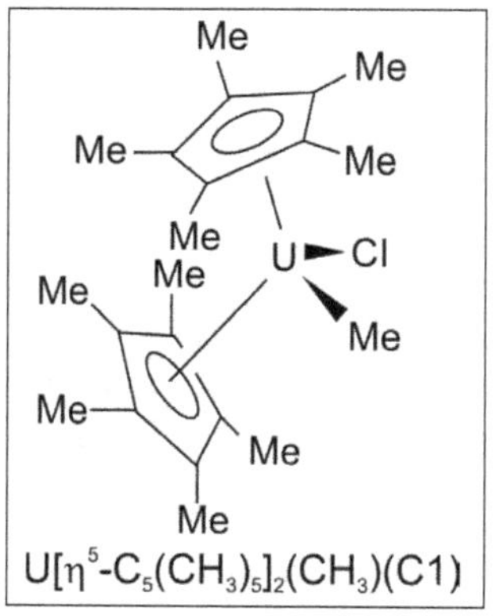

$U[\eta^5\text{-}C_5(CH_3)_5]_2(CH_3)(C1)$

## Metal Clusters

Metal cluster compounds contain metal-metal bonds. The focus here is on compounds having three or more metals in a closed array. Carbon monoxide is the most common ligand in organometallic cluster compounds, but many other organometallic ligands are bound to clusters, and the presence of several metals leads to bonding arrangements for the ligand that are not possible for monometallic compounds. A variety of metal arrays are seen in cluster compounds. Triangular, tetrahedral, and octahedral clusters are common, and much larger metal arrays are known. The structures of many clusters, which can be precisely determined by single-crystal X-ray diffraction, provide some clues to the way in which ligands are bound to the surfaces of bulk metal particles. The latter are more difficult to structurally characterize than are molecular clusters.

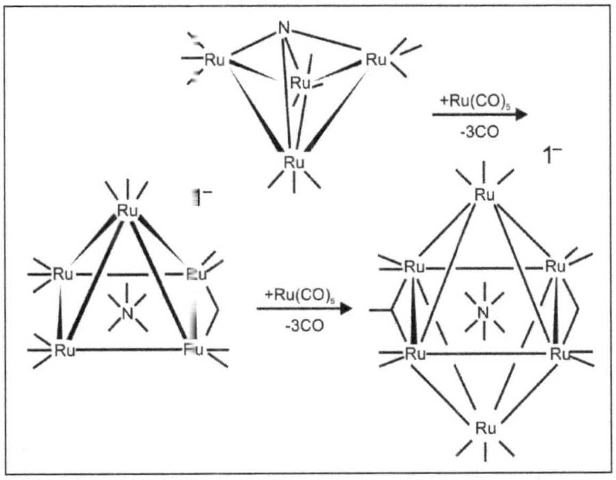


Cluster-building reactionsIn each step, one ruthenium (Ru) metal atom is added to the cluster. The nitrogen (N) atom initially is in an exposed position and is bonded to four Ru atoms. In the intermediate step it is bound to five Ru atoms, and in the final product it sits in the centre of an octahedron of Ru atoms and is bonded to all six of them.

For many d-block clusters there is a strong correlation between their structure and the number of valence electrons (from the metal atoms and the ligands). This set of correlations for clusters is similar to the 18-electron rule for mononuclear organometallics, and these guidelines are often called Wade's rules after the British chemist Kenneth Wade, who first recognized that a triangular cluster such as $Ru_3(CO)_{12}$ usually has 48 valence electrons, a tetrahedron such as $Co_4(CO)_{12}$ has 60 electrons, and an octahedron such as $Rh_6(CO)_{12}(\mu_3\text{-}CO)_4$ has 86 electrons. In some cases, it is possible to synthesize clusters in a stepwise manner. An interesting example of this type is the buildup of a ruthenium nitride cluster; in the process of cluster building, the nitrogen ligand is progressively encapsulated by metal atoms.

## Organometallic Compounds in Catalysis

Catalysts are substances that increase the rate of a reaction but are not consumed in the reaction. Catalysts are widely encountered in nature, industry, and the laboratory. Many of the catalysts utilized in the chemical industry and the laboratory are organometallic compounds.

## Hydrogenation

The overall result of the catalytic hydrogenation of alkenes is to add molecular hydrogen, $H_2$, across the double bond of an alkene. The reactants, $H_2$ and ethylene ($C_2H_4$), enter the cycle by reaction with the complex to produce in succession a hydrido complex and an alkene complex. In the final step, the hydrogenated product leaves the loop with the regeneration of the coordinatively unsaturated Rh complex. The cycle continues as long as hydrogen gas and ethylene are supplied.

The rhodium complexes in solution are the catalysts and are not used up in the reaction. Rhodium, which is more expensive than platinum, can be used even in catalytic processes where the products are inexpensive, because the rhodium is not consumed. Modifications of this type of catalyst are also employed in the production of pharmaceuticals such as levodopa (or L-dopa), which is used to treat Parkinson disease.

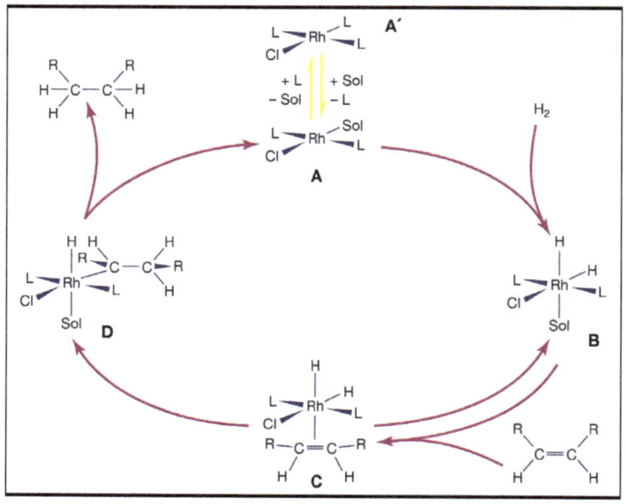


There are multiple steps in the catalytic hydrogenation of alkenes (RHC=CRH compounds) by a rhodium (Rh) complex. L is the triphenylphosphine ligand, $PPh_3$, and Sol is a solvent molecule.

## Hydroformylation

Hydroformylation involves the addition of carbon monoxide and hydrogen to an alkene to form an aldehyde containing one more carbon atom than the original alkene.

$$R{-}CH{=}CH_2 + CO + H_2 \longrightarrow R{-}CH_2{-}CH_2{-}\overset{\overset{\displaystyle O}{\|}}{C}{-}H$$

This catalytic reaction is employed in the petrochemical industry, where $Co_2(CO)_8$ or various rhodium catalysts are utilized. The catalytic cycle proceeds through a series of organometallic intermediates. The aldehydes produced by hydroformylation are normally reduced to alcohols that are used as solvents, as plasticizers, and in the synthesis of detergents. The scale of production is enormous, amounting to millions of tons per year.

## Alkene Polymerization

Polyalkenes, the most common and useful class of synthetic polymers, are often prepared by use of organometallic catalysts, either in solution or supported on a solid surface. In the 1950s, the German chemist Karl Ziegler developed a catalyst for ethylene polymerization based on a catalyst formed by the reaction of $TiCl_4$ with $Al(C_2H_5)_3$. Soon thereafter, Italian chemist Giulio Natta made use of this type of catalyst for the polymerization of propylene to produce polymers with highly regular structures. The intimate details of the reactions of these commercial catalytic processes are not entirely understood, but there are strong indications from more easily studied soluble organometallic catalysts that alkenes coordinate to a metal centre and then insert into a hydrocarbon chain, producing a longer-chain hydrocarbon attached to the metal centre. Repetition of this process leads to extremely-long-chain hydrocarbon polymers, which include many of the most familiar plastics, such as polyethylene and polypropylene. These plastics are used in consumer items ranging from milk containers and plastic bags to artificial limbs and car bumpers.

# Permissions

All chapters in this book are published with permission under the Creative Commons Attribution Share Alike License or equivalent. Every chapter published in this book has been scrutinized by our experts. Their significance has been extensively debated. The topics covered herein carry significant information for a comprehensive understanding. They may even be implemented as practical applications or may be referred to as a beginning point for further studies.

We would like to thank the editorial team for lending their expertise to make the book truly unique. They have played a crucial role in the development of this book. Without their invaluable contributions this book wouldn't have been possible. They have made vital efforts to compile up to date information on the varied aspects of this subject to make this book a valuable addition to the collection of many professionals and students.

This book was conceptualized with the vision of imparting up-to-date and integrated information in this field. To ensure the same, a matchless editorial board was set up. Every individual on the board went through rigorous rounds of assessment to prove their worth. After which they invested a large part of their time researching and compiling the most relevant data for our readers.

The editorial board has been involved in producing this book since its inception. They have spent rigorous hours researching and exploring the diverse topics which have resulted in the successful publishing of this book. They have passed on their knowledge of decades through this book. To expedite this challenging task, the publisher supported the team at every step. A small team of assistant editors was also appointed to further simplify the editing procedure and attain best results for the readers.

Apart from the editorial board, the designing team has also invested a significant amount of their time in understanding the subject and creating the most relevant covers. They scrutinized every image to scout for the most suitable representation of the subject and create an appropriate cover for the book.

The publishing team has been an ardent support to the editorial, designing and production team. Their endless efforts to recruit the best for this project, has resulted in the accomplishment of this book. They are a veteran in the field of academics and their pool of knowledge is as vast as their experience in printing. Their expertise and guidance has proved useful at every step. Their uncompromising quality standards have made this book an exceptional effort. Their encouragement from time to time has been an inspiration for everyone.

The publisher and the editorial board hope that this book will prove to be a valuable piece of knowledge for students, practitioners and scholars across the globe.

# Index

## O



## P



## Q



## R



## S



## T



## U



## V

www.ingramcontent.com/pod-product-compliance
Lightning Source LLC
LaVergne TN
LVHW080652210926
848485LV00003B/367
* 9 7 8 1 6 4 1 7 2 6 3 2 0 *